For those who are interested in how religion and ethical values can make a profound contribution in mitigating the disastrous effects of climate change this book is a compulsory read. It is a thoroughly holistic account of the "why" of climate change and the "how" to do something about it and helps people of faith to relate to the urgent issues of our times. Fazlun Khalid provides an integrated perspective combining faith with environmental, economic and historical narratives thus contributing fresh thinking in these unpredictable times.

**Prof. Dr. Azizan Baharuddin, Director General,
Institute of Islamic Understanding Malaysia**

At last an authoritative statement on this crucial matter from a vital constituency whose voice has, up to now, scarcely been heard at all. Muslims make up a quarter of the world's population and, owing to the areas where they live, are in many cases disproportionately affected by global warming and climate change. Drawing on sources deeply embedded within Islamic tradition, Fazlun Khalid shows that the vast body of knowledge and experience contained within it has much to offer to the contemporary discourse on the ecology of our planet and he provides a perspective that is surprisingly original and extremely relevant to those who care about the future of the world in which we live.

**Shaykh Abdalhaqq Bewley,
Islamic scholar and co-translator with Aisha Bewley of
The Noble Qur'an: A New Rendering of its Meaning in English**

For years, Fazlun Khalid has been the world's leading voice for Muslim environmental action. He has taught us about Islam's beautiful and sacred teachings on the Earth. He has told us, with urgency, to respond. Signs on the Earth is his *magnum opus*. It is electrifying, frightening, challenging, and deeply imbued by decades of reflection and scholarship as a Muslim. As the *azaan* calls the Islamic community to prayer, Fazlun Khalid's voice calls us to action, provokes us to see, feel and act in a manner appropriate to the grave crisis we face. We all need to listen to what he is saying and heed his call to work together.

**The Rev. Fletcher Harper, Executive Director,
GreenFaith, New Jersey, USA**

This book contributes much to an understanding of the holistic nature of the Islamic worldview. The insights it brings to bear sheds much light on the path to discovering solutions to the global environmental crisis.

Professor Odeh Al-Jayyousi, Head of Innovation and Technology Management, Arabian Gulf University, Bahrain. Member of UN GEO6- Scientific Advisory Panel

Climate change and ecological collapse are no longer future theoretical risks, but real and present threats to our security now. If we are to avoid disaster it will be necessary not only to present reason to power, but also to awaken the great faiths. Fazlun Khalid's excellent writing reveals why environmental concerns should be at the heart of Islam, its teachings, its sense of the divine and sacred and its moral and spiritual authority.

Tony Juniper CBE, Executive Director at WWF and former Director of Friends of the Earth

A Great Book indeed! Fazlun Khalid rewards us with the benefit of almost forty years of reflection and activity on the root causes of the formidable environmental problems we have created for ourselves. He also points to where our responses lie drawing inspiration from readings and reflections in the Quran. The book argues that understanding the *Signs on the Earth* and reflecting on their implications will at the very least open the way to perceiving the challenges that face us today in the manner it led the early Muslims to face the challenges of their time. It is a book that moves the senses and certainly deserves careful study especially, but not only, by those who are interested in an Islamic approach to the environment and climate change.

İbrahim Özdemir, author of *The Ethical Dimension of Human Attitude towards Nature: A Muslim Perspective*. Professor of Islamic Philosophy and Theology, Åbo Akademi University, Turku, Finland

Fazlun Khalid has been a leader in Islam and Ecology for over 40 years. His dedication to this is noteworthy as can be seen in his life work. *Signs on the Earth* is a care-fully researched and passionately written plea for the wellbeing of our Earth community. Fazlun's historical analysis and his economic critiques are penetrating and insightful. We hope Fazlun's call to bring forward the rich resources of Islam for a just and flourishing future will be heeded by many."

Prof. John Grim and Prof. Mary Evelyn Tucker Directors, Forum on Religion and Ecology at Yale University

Dr. Fazlun Khalid provides a vivid description of today's complex and complicated world's affairs. Yet, I did not find pessimism but lots of hope. Dr. Fazlun's estimable and vast knowledge of world cultures and history has given the book a different taste and an unmatchable power of message. *Signs on the Earth* is a book of books. I thank Dr. Fazlun for generously offering his knowledge and for reiterating the need for coherence and comprehensiveness in our work integrating religious values and teachings into protecting our only Earth and to prevent what Dr. Fazlun describes as "species-level suicide".

Iyad Abumoghli, Principal Coordinator, Faith for Earth, Strategic Engagement with Faith-based Organizations, Policy and Programme Division, UN Environment Programme.

What a beautiful book and a blessing to read. It brings together so much of Fazlun Khalid's life's work, expressed with passion and eloquence. We will be using it as a key reference in the various aspects of our own work, in biodiversity conservation and ecological ethics, wilderness leadership and environmental education, and environmental design and planning.

Othman Llewellyn, Environmental Planner, Saudi Wildlife Authority, Riyadh, Saudi Arabia

I lack a faith, but hugely value the power of faith in framing issues - this book is a masterclass in how to do that

Fred Pearce, author and environment consultant for New Scientist magazine.

"Signs on the Earth" is a beautifully written work which is the product of a searching mind. It encompasses a critical sweep of history which is impressive.

Joyce D'Silva, D. Litt., D.Univ. (Hons.), Ambassador Emeritus, Compassion in World Farming.

Fazlun Khalid seems to have intended this book as a testament, and one hopes that it will stand as a lasting monument to his legacy. He is a clear thinker and knows how to present a compelling argument. He is equally at home in Western and non-Western contexts and is able to communicate easily across cultures. Readers from all backgrounds will find his presentation informative and inspiring.

Prof. Richard Foltz, Department of Religions and Cultures, Concordia University, Montréal, Canada.

SIGNS ON THE EARTH

ISLAM, MODERNITY AND THE CLIMATE CRISIS

FAZLUN M KHALID

Signs on the Earth: Islam, Modernity and the Climate Crisis

First published in England by
Kube Publishing Ltd
Markfield Conference Centre
Ratby Lane
Markfield
Leicestershire, LE67 9SY
United Kingdom

Tel: +44 (0) 1530 249230
Fax: +44 (0) 1530 249656

Website: www.kubepublishing.com
Email: info@kubepublishing.com

Cover Image: Shutterstock

Cataloguing in-Publication Data is available from the British Library

ISBN 978-1-84774-076-2 Casebound
ISBN 978-1-84774-075-5 Paperback
ISBN 978-1-84774-077-9 ebook

Cover Design: Inspiral Design
Typesetting: LiteBook Prepress Services
Printed by: Clays Ltd, UK

The inspiration for the title of this book was drawn from the following verses in the Qur'an:

There are signs on the Earth for people with certainty.
And in your selves as well. Do you not then see?
(51: 20–21)[1]

To my daughter Shireen Laila – for always a child
and
All the children of the world

CONTENTS

FOREWORD AND
ACKNOWLEDGEMENTS

Alarm bells are ringing, presaging an irreversible human-induced climate change threatening the world-wide collapse of ecosystems. Previously ignored and marginalized, faith communities are now emerging as significant contributors to the restoration and protection of the biosphere. This volume is a contribution to this process from an Islamic perspective. It has been gestating for nigh on forty years and its birth pangs have lasted five. Looking back, the world in the pre-foetal stage of the book was of a different order.

I am not alone in this journey, but the need is urgent and requires a rapid response. It entails lifting our consciousness to another level of reality. The main challenge for us is to alert and mobilize the billion and a half strong community of Muslims to the responses that are needed to restore the Earth to a state of equilibrium. But how to do this? One original idea was to create an eco-community of Muslims that would serve as a template for people looking for a conservation model based on Islam. This idea had to be shelved due to shortages of resources, although there is still hope that it can be resurrected. I established the Islamic Foundation for Ecology and Environmental Sciences (IFEES/*EcoIslam*) in 1994 as a charitable trust and was fortunate enough to have a board of trustees who saw the importance of the work, gave it their unstinting support and enriched it by their own knowledge and personal experience. My thanks to the following members of the Board for their support and for giving me the encouragement and guidance I needed to both develop the work and write this book: Ashfaq Ahmed, Harfiyah Abdelhaleem, Shaid Latif, Ejaz Qureshi, Lutfi Radwan and Ahmad Thomson. It would be remiss of me not to mention past members who have given similar support, and they are Tareq Ali, Adnan Ashfaq, Fatima D'Oyen, Muhammad Imran, Idris Mears, Tasib Mughal, Fuad Nahdi and Faisal Osman.

IFEES is essentially an educational organization, and the aspirations behind it were to gather texts from the Qur'an and hadith literature, formulate them into a model that would be understood today and establish conservation projects based on this teaching. Much of the material in this book, particularly chapter 5, is the result of this work. Looking back on it, and given the recognition IFEES has received internationally over the years, it appears to have worked but this could only have happened with the encouragement and support of the following, both past and present, to whom I am grateful: Yusuf Adams, Iftikhar Awan, Hani al Banna, Mark Bryant, Sami Bryant, Saffet Catovic, Hashim Dockrat, Anwar Fazal, Nana Firman, Kamran Fazil, Kate Fryer, Muzammal Hussain, Shaid Hussain, Mohammed Khalid, Zafar Khalid, Ali Khan, Abdur-Razzaq Lubis, Zulkifli Lubis, Ziad Lorgat, Mahmud Lund, Usman Modibbo, Shabaaz Moghal, Khoo Salma Nasution, Dawud Price, Rancho Prime, Aslam Qureshi, Mizan Raja, Lorna Slade, Muhammad Junaid Smith, Ali Khamis Thani, Rianne C ten Veen, Rob Wild; also Samina Faiz and Shagufta Yaqub for their pioneering journalism in producing *EcoIslam* and to Musab Bora for his technical inputs. I would also like to register my thanks to Yusuf Islam, The Islamic Foundation, Muslim Education Consultative Committee, Muslim Hands, Islamic Help and Islamic Relief Worldwide for their support during the course of the development of this work.

The wide-ranging scope of this book has been made possible by the influence of scholars and activists who are numerous. My thanks are due particularly to Khurram Bashir, Shaykh Abdalhaqq Bewley, Mawil Izzi Dien, Prof. Yasin Dutton, Prof. John Grim, Prof. Richard Foltz, Prof. Sophie Gilliat-Ray, Fletcher Harper, Prof. Muhammed Hyder, Prof. Seyyed Hossein Nasr, Prof. Jørgen Nielsen, Shaykh Abdullah Omar Nasseef, Joan O'Brien, Abdalhalim Orr, Martin Palmer, Mohammad Aslam Parvaiz, Shaykh Abdalqadir as-Sufi (aka Ian Dallas), Muhammad Ibrahim Surty and Prof. Mary Evelyn Tucker. My thanks are also due to Prof. Azizan Baharuddin, Othman Llewellyn, Fachruddin Mangunjaya, Prof. İbrahim Özdemir and Abdelmajid Tribak who came together to produce the Islamic Declaration on Global Climate Change and for Muhtari Aminu-Kano and Lotifa Begum of Islamic Relief Worldwide for their invaluable background input that made this happen. I am indebted to Abdassamad Clarke, Tony Juniper, Muzammal Hussain, Harfiyah Abdelhaleem Uthman Ibrahim-Morrison for their thoughts and suggestions on improving certain sections of the draft; also Ejaz Qureshi for producing the diagrams and tables. My thanks are also due to the editors at Kube for staying the course with me; Yahya Birt for the initial discussions that led to the present shape of the book, Siddique Seddon for his editorial assistance, Usaama al-Azami, who took much of the responsibility onto his shoulders and whose suggestions and support were invaluable and last but not least to Keith Devereux, for the additional research and his assistance in ironing out the wrinkles in the text.

I am particularly grateful for the publishers of (a) *Living Lightly, Living Faithfully: Religious Faiths and the Future of Sustainability*, edited by Colin Bell, Jonathan Chaplin and Robert White; The Faraday Institute for Science and Religion, St Edmund's College and the Kirby Laing Institute for Christian Ethics, Cambridge, for the use of material from my essay entitled 'The Environment and Sustainability: An Islamic perspective', chapter 14, and (b) *Religion and Ecology*, edited by John Hart; Wiley Blackwell, Oxford, for the use of material from my essay entitled 'Exploring Environmental Ethics in Islam: Insights from the Qur'an and the Practice of Prophet Muhammad', chapter 11.

It is not uncommon for writers to thank their spouses, without whose support endeavours of this nature would be almost impossible. Writing is a lonely experience, as others who have been through this exercise will confirm. My grateful thanks are due to my wife, Saba, for the enduring patience and support she has given me during what has been a prolonged period of reading and writing, and for the exchange of ideas and periods of intense debate and discussion between ourselves. In addition to thanking her, the least I could do is to give her the same consideration that she has given me as she finishes her own book, which hopefully will be out before long.

My apologies to those whose names I have failed to mention, as this has been a long journey resulting in many a stimulating encounter. I accept responsibility for all errors and omissions in the book as they are entirely mine.

Fazlun Khalid
Birmingham, England
June 2018

INTRODUCTION

On its maiden voyage more than a hundred years ago, when the *Titanic* collided with an iceberg in the North Atlantic the passengers hardly felt it. Less than three hours later the great ship lay at the bottom of the ocean and of the 2,240 passengers and crew more than 1,500 perished. The rest were able to escape into the cold ocean in lifeboats, but if the lookout hadn't warned of the iceberg ahead there would have been a head-on collision, not a glancing blow as it so happened, and the ship would have sunk far quicker, resulting in more fatalities. Given the information we already have of the voyage of good ship Earth, it is no exaggeration to say that it has already hit the iceberg. For now it looks like a glancing blow and the lookouts, in our case scientists and academics, voice their concerns using the kind of emotive language we normally associate with less informed people: uncertain futures; collapsing ecosystems; the end of nature; a tipping point; the end of civilization; a doomsday scenario; the end of humankind, collapse or extinction, and so on. But our responses so far have been akin to repositioning deckchairs on the *Titanic*, and even if we have lifeboats there is nowhere else we could go. The point is we still have time to change, even if it is only a small window of opportunity.

The good ship Earth has been voyaging for millennia through storms and squalls, fog and mist and thick and thin and has managed to keep afloat. Then, in more recent times, it changed course and steered in the direction we now call modernity, although nobody knew where it was heading. Where then are we heading? What is modernity? Is it about leading comfortable lives filled with luxury, leisure, holidays, consumerism, push-button civilization, cars, high-speed travel, literature, arts, music and the theatre? Or is it about economics, progress, development and prosperity? Or secularism, liberalism, and individualism? Or is it about science, technology, social media, artificial intelligence or robotics? Or changing values, favouring the new over the old, denying the authority of the past, looking at tradition irreverently as old, obsolete, antiquated, as something that should be abandoned and replaced?[2] The philosopher John Gray observes:

> At bottom the modern world is a jumble of things engendered by the accelerating advance of knowledge. The spread of literacy and the growth of cities, the expansion of trade and the spread of industry all these trends are by-products of expanding scientific knowledge. None works to promote any particular values.[3]

So values are not the concern of modernity but outcomes are, and they are quantifiable. It is about being bigger and better and reveals itself in GDP, GNP, economic growth, the development index, the happiness index, per capita incomes, national debt, bank balances, interest rates, stock market indices, tall buildings, highways stretching to the horizon, mega dams, and traffic jams. It is also about grandeur, the insatiable ego, the display of power and ostentation, getting as close to the top of the pile as one can get, and bigger and better is there to show for it. As modernity advanced the divine and the sacred receded from our lives. As the world became increasingly secular, faith receded into ritual, an occasional anchor in difficult times. Modernity has also given some of us comfort, good health, education and a greater understanding of the world we live in, but where is all this taking us? The Qur'an observes:

> *As for the Earth We have spread it out wide*
> *And set upon it firmly embedded mountains*
> *And made everything grow there in balance.*
> *And We have provided means of sustenance for you*
> *And for all those creatures who do not depend on you.*
> *There is not a thing whose storehouses are not with Us*
> *And We only send it down in appropriate measure.*
> *We send forth pollinating winds*
> *And bring down water from the sky for you to drink*
> *And you do not control its sources.* (15: 19–22)

The natural world is now a resource to be exploited not an entity to be cherished; a theatre for our follies. So how do we overcome insurmountable odds in dealing with the environmental debacle we have created? An optimistic climate control agreement was reached in Paris in 2015 but the US, the world's greatest polluter, has chosen to withdraw from it. A successful agreement on Sustainable Development was concluded by the international community the same year, but nation states continue with programmes of unsustainable economic growth. There is very little evidence that environmentalists and economists are talking to each other about these overarching global issues. The World Bank and IMF encourage economic growth whilst United Nations agencies worry about depleting rain forests, over fishing, species extinction and a plethora of environmental concerns. An assessment of the environmental impact of global conflict, civil war and the refugee pandemic is conspicuous by its absence. The arms industry thrives, profiting from peddling death. Then there is the nuclear stockpile which can unleash Armageddon at any moment. We also have globalization, and Joseph Stiglitz, the one-time chief economist of the World Bank, observes:

> ... the benefits of globalization have been less than its advocates claim, the price paid has been greater, as the environment has been destroyed, as political processes have been corrupted, and as the rapid pace of change has not allowed countries time for cultural adaptation. The crises that have brought in their wake massive unemployment have, in turn, being followed by longer-term problems of social dissolution ...[4]

There is a manifest lack of an integrated narrative as the UN is reduced to a showcase for jingoism.

An economics text book I was reading in the early 1960s, during my pre-university studies, had something very strange to say about banking and money, and it was this: If I asked my bank for a loan and it was approved I would be able to withdraw money from the bank up to the value of my loan, although the bank did not have any money to give. This had me puzzled, if only for a while, and it didn't take long for this piece of information to subside to the bottom recesses of my mind as life's mundane urgencies distracted me. This conundrum resurfaced again in the mid-1980s, when I attended a workshop on the subject of currency and banking set up by a group of European converts to Islam.[5] I had by this time become politicized, not a difficult transition for an ex-colonial like myself, helped along by a busy interlude as a trade unionist and a twenty-three year stint with the Race Relations Board and the Commission for Racial Equality, the predecessor bodies of the Equality and Human Rights Commission. I had also developed a more than passing interest in so called 'Third World' issues concerning hunger and debt during this period and found them morphing into a global movement addressing environmental concerns during the 1960s and 70s. These activities led me to the discovery of a fundamental connection between the nature of the money that feeds our current civilization and its ultimate expression, which is responsible for the environmental debacle that is advancing upon us, and predicted to get a lot worse. It is the greatest confidence trick of all time, resulting in a consensus amongst the scientific community that the human species has now changed the course of nature.

The global banking crisis that culminated in 2008 let the cat out of the bag. It not only exposed the fragility of the system, it also demonstrated the depths to which banks would descend to shore up their profits. In the end the banks were bailed out by taxpayers and as Mervin King, the ex-Governor of the Bank of England, observed, 'For the banks, it was ... heads I win, tails you – the taxpayer – lose. ... All banks, and large ones in particular, benefited from an implicit taxpayer guarantee'.[6] The Qur'an is quie trenchant about the usury/interest (the difference is merely academic) the banking system thrives on, and it warns, '*forego any outstanding dues from usury (interest) ... if you do not then be warned of war from Allah and His Messenger*'. (2: 278–279) The financial crisis forced governments of the so-called developed world to print money worth trillions of dollars to keep their economies moving. This activity was euphemistically described as 'quantitative easing' (QE).

Rapid population and economic growth are taxing an overloaded, over heated planet that is being brought to its knees by an economic model that conjures money out of thin air. New thinking requires that we re-examine what we have wrought in the name of progress and development, and we redefine the idea of prosperity bearing in mind the finite nature of planet Earth. As the planet degrades, the gap between the haves and the have-nots widens. Inter-generational justice calls urgently for a shift away from the political economic model that has caused so much destruction to one that ensures the natural world remains in balance, so that it can continue to work for us in the manner that it has always done. If we are to continue to benefit from the gifts of the natural world we need to establish a political economy that recognizes Earth's finite resources, ensures distributive justice and re-establishes organic social structures compatible with natural planetary systems: a paradigm shift more in tune with the rhythms of the natural world, not a society based on the profit margins of big corporations and stock market share indices.

This book is an attempt to discover the relationship between modernity and Islam in the context of the environmental concerns which now engage our attention. It looks at how modernity has imposed its stamp not only on the Islamic world but also on all other traditions and ways of life. It examines how this has unfolded in the times in which we live and has succeeded in corrupting all natural systems that support life, while holding out a promise of progress towards a better future.

The ethos of Islam is that it integrates belief with a code of conduct which pays heed to the essence of the natural world:

> It has often been observed that Islam cannot ordinarily be described as a religion and that it prescribes a way of life that goes beyond the performance of rituals ... It provides a holistic approach to existence, it does not differentiate between the sacred and the secular and neither does it place a distinction between the world of humankind and the world of nature ... '*The creation of the heavens and the earth is far greater than the creation of humankind. But most of humankind do not know it*' (40: 57).[7]

In exploring how Islam can contribute to the task of restoring the natural world to a state of equilibrium I also recognize that other traditions will also have a say in the matter. This affair can only be resolved by recognizing our collective responsibility for creating the problem in the first place. People of faith and non-faith traditions are already working together on these issues, and a concerted effort to further strengthen this partnership is essential if we are to leave an Earth which future generations will find inheriting worthwhile.

Endnotes

1. To capture the nuances of the Qur'an, I have consulted the following translations: (a) Abdullah Yusuf Ali, *The Holy Quran*, a standard work first published in 1934 that is commonly available in numerous imprints; (b) A J Arberry, *The Koran Interpreted* (London: George Allen and Unwin, 1955), recognized for its rhythm and consistency; (c) Muhammad Asad, *The Message of The Quran* (Gibraltar: Dar Al Andalus, 1980) gives access to wider sources: (d) A and A Bewley, *The Noble Qur'an* (Norwich: Bookwork, 1999). A very readable modern translation, but watch out for the slight modifications to the numbering system; (e) M A S Abdel Haleem, *The Qur'an* (Oxford: Oxford University Press, 2004), another modern translation; (f) Mohammed Marmaduke Pickthall, *The Meaning of the Glorious Qur'an*, (Hyderabad, Deccan, India: Government Central Press, 1938), another classic translation alongside Yusuf Ali's. The numbering system in the Yusuf Ali translation is used when quoting references to the Qur'an. The chapter and verse numbers are given in parentheses. The names of the chapters have been left out as there are minor variations from one translation to another.

2. Zygmunt Bauman, 'Modernity' in Joel Krieger (ed.), *The Oxford Companion to Politics of the World* (New York: Oxford University Press, 1993), pp. 592–596.

3. John Gray, *Al Qaeda and What It Means to be Modern* (London: Faber and Faber, 2003), p. 108.

4. Joseph Stiglitz, *Globalization and its Discontents* (London: Penguin Books, 2002), p. 8.

5. Abdalhalim Orr and Abdassamad Clarke (eds), *Banking: The Root Causes of the Injustices of Our Time*, revised edition (London: Diwan Press, 2009).

6. Mervyn King, *The End of Alchemy: Money, Banking and the Future of the Global Economy* (London: Little Brown, 2016), p. 96.

7. Fazlun M Khalid, 'Islam and the Environment', in Peter Timmerman and Ted Munn (eds), *Encyclopaedia of Global Environmental Change, volume 5: Social and economic dimensions of global environmental change* (Chichester: John Wiley & Sons, 2002), pp. 332–339.

1

A HOUSE OF CARDS

Hegemony

Encirclement

When Genghis Khan and his descendants charged west into Europe and south into China, leading their horsemen into battle in the twelfth and thirteenth centuries, they perfected the use of the stirrup and changed the nature of warfare in their time. High speed mobile combat was born as the swift ponies of the Mongols provided mobility and the stirrup provided balance and manoeuvrability. Galloping at speed and swivelling on their hips, the Mongols used their newly freed hands to let loose a rapid succession of arrows against their enemies. A comparable innovation, 'The abandonment of the oar for propulsion and mounting, broadside, of large numbers of guns ...'[1] enabled the Europeans to control the seas from the fifteenth century onwards. As sails displaced oarsmen, who slipped into the role of gunners, a small number of ships were now able to control the vast oceans and the Europeans used fear, together with a brutality that was matched only by the Mongols just two centuries before them, to create an empire. Gunboat diplomacy had arrived and with that a global hegemony of one group of people over all others, which thrives to this day. This is reflected in the control and dominance of the European/Western powers since the fifteenth century over all others in the political, economic, social and cultural lives of the people of the world. Although geopolitics is changing, American fleets still circle the globe.

There was, however, an essential difference between the Mongolian mindset and that of the European. The former were not so much interested in the spread of their culture – for it would appear they were liberal in this regard and welcomed missionaries of all persuasions – as they were in holding onto power, whereas for the Europeans power became a means of transferring wealth to the home countries on a colossal scale, as well as imposing a form of cultural and economic domination that would forever change people's relationship to their natural environment and their own local traditions. From its very onset, European exploration was impregnated with a crusading and missionary zeal which in its expression bore little or no relationship to the core teachings of Christianity. Prince Henry the Navigator of Portugal, who controlled marine exploration in the middle of the fifteenth century on behalf of his brother king, is said to have been motivated by a set of objectives which came to be accepted by explorers after him. They were amongst other things to discover new lands, to trade with Christians, win converts to Christianity and to discover the extent of Islamic power.[2] Monarchs, explorers and adventurers received further encouragement from the Papal Bull promulgated on 4 May 1493 by Pope Alexander VI that provided justification for conquest and conversion. The New York based Centre for Earth Ethics has made representations to Pope Francis to 'rescind the Papal Bulls of the Fifteenth Century that gave theological justification to the murder, enslavement and oppression of the peoples of the Americas and Africa'.[3]

In the fifteenth and sixteenth centuries the Catholic Iberians had the oceans virtually to themselves: for the Spanish it was the Atlantic and for the Portuguese the Indian Ocean theatre right up to the rim of the Pacific Ocean all the way to China. While Spaniards were carving out an Empire in South America, they also managed to cobble the Philippines together and hold them for more than three centuries, from the sixteenth to the end of the nineteenth. The Portuguese, in addition to claiming the Brazilian slice of the South American cake as their own, were also building trading posts in the East. This monopoly was broken as the seventeenth century was approaching, when Protestant England and the Netherlands began to make inroads into this Iberian hegemony. An interesting fact about this enterprise at this stage of history was that the Orientals were more interested in the gold and silver they received in exchange for their products rather than anything else the Europeans had to offer. 'When Vasco da Gama showed what he brought to give to a king, the inhabitants of Calicut laughed at him; he had nothing to offer which could compare with what Arab traders already brought to India from other parts of Asia.'[4] The Europeans for their part resented having to part with gold and silver for the merchandise the Orientals had to offer. European trading stations became fortresses, which in turn evolved into mini colonies defended by garrisons resulting in infantry and artillery being imported from the mother countries. They were partly a protection from the locals but mostly a defence against marauding European rivals; their cannons nearly always pointed out to sea. The first phase of imperialism had arrived and the logic was quite simple: If you colonized the land you owned the produce. Cut out the middle-men and all the profits were yours and the gold and silver remained in your hands. Local rulers manipulated European rivals to gain relief, and a good example of this was how the Sinhalese kings of Ceylon (present day Sri Lanka) pitted the Dutch against the Portuguese in the 1630s and 40s. This backfired in 1645, when the Europeans formed an alliance to plunder the cinnamon that grew so profusely in the coastal regions. Greed took precedence over any promises made in a treaty and there was more of this to come.

The Spanish destruction of South American civilizations is well known, but the Dutch – who like to think of themselves as more humane and tolerant in their colonial adventures – have a few skeletons in their cupboards too. 'The early history of the Dutch colonies is a grim one of insurrection, deportation, enslavement and extermination. The trade of local shippers – and of Chinese junks – was deliberately destroyed in order to concentrate all sources of profit into the hands of the Dutch.'[5] In 1740, 10,000 ethnic Chinese were massacred in a pogrom in Batavia (present day Jakarta).[6] The treatment of the Acehenese in Northern Sumatra towards the end of the nineteenth century, who the Dutch could not bring under their yoke, is not something they can be proud of today, nor of their actions in the closing years of the Second World War in the eastern theatre, when Indonesians were fighting for their independence.

'The interchange of trading goods, disease pools, and ideologies that took place from 1500 to 1700 created the foundations of the world we inhabit today.'[7] The idea of the interchange of trading goods is misleading, as hinted earlier, because in the period alluded to the movement of the vast bulk of produce and merchandise was centripetal, in the sense that Europe was at its centre. This 'interchange' included plunder, as the wealth of the world was sucked into Europe in vast quantities. Disease pools, ideologies and the Christian missionary movement were centrifugal in the sense that these entities radiated out of Europe. But in the end these movements did have the effect of creating the world we live in today. But at what cost! These processes set in motion the wholesale destruction of civilizations, cultures, traditional communities and the natural world on a scale previously unknown.

The saga of European global encirclement will not be complete without mention of the French and the United States of America. The French belatedly discovered that there were some portions of the far eastern cake still to be had and as a result homed in on Vietnam with a classic display of gunboat diplomacy. By 1859 they had established sufficiently strong bridgeheads in the country, and by 1887 the French were masters of what is now Vietnam, Cambodia and Laos. They imposed humiliating conditions on the rulers of these countries with the sole 'aim [of] rapid and systematic exploitation of Indochina's (as this region came to be known) potential wealth for the benefit of France; Vietnam was to become a source of raw materials and a market for tariff-protected goods produced by French industries. The exploitation of natural resources for direct export was the chief purpose of all French investments'.[8]

The encirclement of the globe was complete when a fleet of ships from the United States of America, led by Commodore Matthew Perry, sailed into Tokyo harbour on 8 July 1853. 'Perry, on behalf of the U.S. government, forced Japan to enter into trade with the United States and demanded a treaty permitting trade and the opening of Japanese ports to U.S. merchant ships.'[9] Although the American version of gunboat diplomacy came much later in the day, they were merely following the example set by their European counterparts. It could be argued that these nineteenth century events, particularly the emergence of the US as an imperial power, set in train the movement we understand today as globalization. This was also the time when the terms 'European' and 'Western' came to be used interchangeably.[10]

The historian Carroll Quigley estimates that about twenty major civilizations existed in all of human history; civilization here meaning a society fixed in time with well developed social organizations, an advanced culture and an organized way of life. He lists sixteen of these civilizations and estimates that:

> Of these sixteen, twelve, possibly fourteen, are already dead or dying, their cultures destroyed by outsiders able to come in with sufficient power to disrupt the civilization, destroy its established modes of thought and action, and eventually wipe it out. Of these twelve dead or dying cultures, six have been destroyed by Europeans bearing the culture of Western Civilization.[11]

The six civilizations that were destroyed since the European sea voyages began were the Andean, Mayan, Hindu, Chinese, Japanese and Islamic. 'When we consider the untold numbers of other societies, simpler than civilizations, which Western Civilization has destroyed or is now destroying, societies such as the Hottentots, the Iroquois, the Tasmanians, the Navahos, the Caribs, and countless others, the full frightening power of Western Civilization becomes obvious.'[12] According to Quigley's estimates it took six and a half millennia from 6000 BCE for fourteen or so succeeding civilizations to destroy each other up to the European period. The Europeans managed to destroy six thriving civilizations since the start of their voyages in a matter of just five hundred years.

An example of nineteenth century imperial attitudes behind this destruction can be gleaned from the works of the Earl of Cromer, who wrote in a personal history of Modern Egypt: 'It is absurd to suppose that Europe will look on as a passive spectator whilst a retrograde government, based on purely Mohammedan principles and obsolete oriental ideas, is established in Egypt.'[13]

Consolidation

The final decade of the fifteenth century saw Muslims and Jews expelled from the Iberian Peninsula. That they were native Iberian themselves and that they were there since the eighth century seems to have mattered little. (This is one of the backwaters of history and bears comparison to the great population upheavals when India was partitioned by the British in 1948.) However, as the sixteenth century dawned the Iberians were thrusting out into the oceans west and east. Columbus had already crossed the Atlantic and Vasco Da Gama had rounded the Cape of Good Hope. This signalled an unprecedented change of direction in world history and 'never before had one culture spread over the whole globe'[14] in this manner. Prior to the European incursions the world had always been culturally multi-polar and the historical tendency had been towards differentiation. But since 1500 there was an overwhelming onslaught that ensured an irreversible swing towards hegemonic political, economic and cultural domination. European settler populations were beginning to occupy the whole of the Americas from the beginning of the fifteenth century, and by the end of the eighteenth they had begun to establish themselves in Southern Africa and the distant antipodes. This was not merely a transfer of people, it was also the beginnings of the establishment of European cultural outposts from one corner of the world to the other. In the view of the historian JM Roberts, Christianity was a 'psychological asset'[15] in the hands of the settlers, and it was this soft power that finally tamed the natives:

> ... this found vent in missionary enterprises, but it was always present as a cultural fact, ensuring the European of his superiority to the peoples with whom he began to come into contact for the first time. [This was] often to have disastrous effects. Confident in the possession of the true religion, Europeans were impatient and contemptuous of the values and achievements of the civilizations they disturbed. The result was always uncomfortable and often brutal ... blur[ring] easily into less avowable motives.[16]

There were two different recognizable strategies of the colonizing process; one followed by the Catholic Church, supported by the Papal Bull of 1493,[17] and the other by the Protestants. What they had in common was missionary zeal, with the former under papal authority committed to domination by the church and the latter more measured in its approach with an eye on profit margins. 'They (the Catholic Church) took (native American) Indians from their tribes and villages, taught them Christianity and Latin ... put them in trousers and sent them back to spread the light among their compatriots.'[18] This is not a great deal different to what is happening in Africa today. The nation states that eventually emerged in South America were based on the locations of the mission stations. But before all this was to happen the heart of the old culture had to be destroyed, and the Spanish decimation of South American civilizations is considered to be one of the most brutal in European colonial history: Pizarro's treatment of the Inca King Atahualpa was 'one of the most atrocious acts of perfidy on the record of history!'[19]

The Protestants had a different strategy, and the historian Roberts, drawing from the imperial experience, observes that the British initially 'lacked missionary zeal ... [and] Protestant interest in missions quickened later than Catholic, and ... had no wish to interfere with native custom or institution but only ... to provide a neutral structure of power ... while the commerce from which the company [British East India Company] profited prospered in peace.'[20] Monarch, church and the conquering army did not have the tight Trinity-like structure of the Iberians. The British and Dutch were literally business like, as exemplified by the Dutch East India Company and its British counterpart. However, their attitudes towards the conquered peoples were exemplified by a mixture of unbridled arrogance and racist paternalism. Although conversion did not merit the same priority as the Catholics, they did not hesitate to kidnap native children from their parents and incarcerate them in mission schools, as happened with the people native to North America and Australia. They sought to change people to resemble themselves but to carry on being 'native' if they wished to – but at a cost to themselves. Ample evidence of this can be found today in reservations in North America and Australia, where the descendants of the original inhabitants of these countries continue to lead pitiful lives.

Some clues to these attitudes can be gleaned from the 'Minutes on Indian Education' produced by Thomas Babington Macaulay, member of the British parliament, in the debates leading up to the 1835 English Education Act (of India). On the one hand he asserts that he could not find any distinguished (European) scholars 'who could deny that a single shelf of a good European library was worth the whole native literature of India and Arabia',[21] and one is not certain whether this is crass political rhetoric or just plain ignorance but the arrogance is plain for all to see, whereas on the other hand a sense of the paternal emerges when he asserts 'that it is possible to make natives of this country thoroughly good English scholars, and that to this end our efforts ought to be directed.'[22] It was in no sense intended that these good English scholars would be free English scholars but that was to come over a hundred years later, and not without a struggle, by which time English would become one of the numerous languages of India. Macaulay's prescription for the interim was to:

> ... do our best to form a class who may be interpreters between us and the millions whom we govern, a class of persons, Indian in blood and colour, but English in taste, in opinions, in morals and in intellect. To that class we may leave it to refine the vernacular dialects of the country, to enrich those dialects with terms of science borrowed from the Western nomenclature, and to render them by degrees fit vehicles for conveying knowledge to the great mass of the population.[23]

The Catholics and Protestants applied different means but the end was the same.

It was Macaulay's oratory in his subsequent role as Secretary of State for War that led to the opium wars against China and that country's humiliation.[24] This was the matter of the British trying to force the Chinese to import the opium they did not want with the persuasive powers of the gunboat. Imperial control was established by massive programmes of social engineering supported by the soft power of the missionary networks, as in the English Education Act in India, or violence, as in the opium wars in China. As the curtain was falling on the imperial period this was turned into a fine art by the British and the French. How the Ottoman Empire was dismantled would provide us with some clues as to how subterfuge, broken promises and unashamed trickery would serve the imperial cause, a cause that has drawn us all into the Middle East quagmire which the entire world has come to witness.

Humiliation

As Ottoman Turkey was being gnawed away at the edges, inventive ways of bringing a civilization to its knees without a match being lit to fire a cannon were being discovered: a new weapon of mass destruction was being put together. In the second half of the nineteenth century, the ambitious Khedive Ismail of Egypt was led by a fraternity of British banks to create a mountain of debt which the country couldn't possibly repay. The Khedive was subjected to punitive and shameful repayment conditions to a loan (one of many) of £32 million he negotiated with the Rothschilds' investment bank in 1873:

> The Rothschilds kept nearly £12 million as security and, of the 20 million actually handed over, some £9 million of it was in substantially overvalued bonds of Egyptian floating debt. The Egyptians received less than half of what they had borrowed and, of course, had to pay interest on the whole of it. This was fraud on a massive scale that goes unmarked by most historians. The 1873 loan, instead of alleviating the Egyptian position seriously weakened it. Bankruptcy was only avoided in 1875 by the sale of the Egyptian government's share in the Suez Canal to the British government for a derisory £4 million.[25]

From this point onwards the Egyptian government was for all intents and purposes bankrupt. 'Egypt was delivered into the hands of European financial interests, who, with the support of the British and French governments, progressively took over the running of the country'.[26] Direct political control was established when these governments appointed their nationals to take over the Finance and Public Works ministries in 1878.

The Europeans also had the capacity to fight a brutal war amongst themselves and at the same time pounce on the Ottoman caliphate like a pack of wolves savaging an ageing bear. During the course of the First World War, the Russians were nibbling away at the Ottoman's eastern flank while the British and French were carving up the western with the help of the Arabs. While one part of the British government based in London was encouraging Sharif Hussein of Mecca to revolt against the Turks on the promise of sovereignty over a large slice of the Arab lands, another part based in Delhi was negotiating with Ibn Saud of the Nejd region, which ended with an agreement that recognized his independence from Sharif Hussein. Thus emerged two sheikhdoms, Hejazi and Nejdi, in the heart of Islam created and backed by the British. The loser in all this was Sharif Hussein. The notorious Sykes–Picot Agreement (the names of the British and French diplomats, respectively, responsible for drawing up the agreement) carved up the Middle East as the Ottomans were collapsing, in total disregard to the promises made to Sharif Hussein. He was finally defeated by the Saudis, who were to take over a rump of the land promised to the Arabs. 'There is a sharp contrast between the imperialist avarice to be found in the secret agreements like Sykes–Picot and the altruistic tone of the publicly issued statements.'[27]

Quigley observes with regard to the Balfour Declaration, which is seen to be the origin of the creation of the state of Israel:

... that this was neither an agreement nor a promise but merely a unilateral declaration (contained in a letter to Lord Rothschild), that it did not promise a Jewish State in Palestine or even Palestine as a home for the Jews, but merely proposed such a home in Palestine and that it reserved certain rights for the existing groups in the area. (Sharif) Hussein was so distressed when he heard of it that he asked for an explanation, and was assured by D.G. Hogarth, on behalf of the British government, that 'Jewish settlement in Palestine would only be allowed in so far as would be consistent with the political and economic freedom of the Arab population.'[28]

The Arab population (Palestinians) are still waiting and Quigley further observes that, 'there is also a sharp contrast between the tenor of the British negotiations with the Jews and those with the Arabs regarding the disposition of Palestine'.[29] There are now two Palestinian territories in the Middle East under the very nose of Europe – the West Bank and the Gaza Strip – where the Palestinians have been holed up since 1948 after they were ethnically cleansed by the Zionist forces that created the state of Israel.

Having dismantled the Ottoman caliphate (the centenary of this event is in 2023), created artificial states in the Middle East that suited Franco-British interests – where future prospects for peace between the Israelis and Palestinians are as remote as ever, especially after the clumsy American action in recognizing Jerusalem as the capital of Israel – supported Iraq in its war against Iran in the 1980s and subsequently attacked Iraq on two occasions, in 1990 and 2003, the West now wonders what it is to do with the cauldron of hate that its military adventurism continues to spawn. Russia, with its own aspirations of global power, has now joined the fray, stoking more resentment and compounding a situation already considered beyond repair, and successfully ensuring that the Syrian regime continues to slaughter its own people. Old habits die hard, and the power play continues at the expense of the weak, which is inevitably seen by the Muslim masses as Christian states continuing their centuries-long interference in their affairs. To this I must add the Iranian adventure in this theatre, further stoking the flames of sectarian conflict.

All this led to the birth of a violent ideology emerging from the rubble and calling itself 'Islamic State' (sometimes referred to as Daesh). Although Daesh has now been all but wiped out, if only in the geographical sense in war-torn Syria, it is now scattered into pockets, creating a state of asymmetric global warfare and generating tensions in western capital cities. Daesh's appalling treatment of Yazidi and Christian minorities – and its prisoners – is beyond comprehension and they bear little resemblance to my understanding of Islam, the faith I was born into and have been practising all my life.

We now live in a world where it comes as no surprise to anyone to see images of an American president on our TV screens, sitting in the comfort of the White House in Washington, monitoring his helicopter gunships mounting an attack on a fugitive in Pakistan.[30] Imperial power remains to this day. Hegemony is alive and well and continues to rule, and we shall uncover more of this before the end of this chapter. Since this book is about our relationships with the natural world, we need to reflect on how all this conflict impacts on the environment. Ban Ki-moon, when he was Secretary-General of the United Nations, had this to say in 2014 in a statement for the UN's International Day for Preventing the Exploitation of the Environment in War and Armed Conflict:[31] 'The environment has long been a silent casualty of war and armed conflict. From the contamination of land and the destruction of forests to the plunder of natural resources and the collapse of management systems, the environmental consequences of war are often widespread and devastating'.

Brains, black holes and the Enlightenment

Knowledge overload

Cosmologists tell us that a black hole is a location in space in which the force of gravity is so strong that not even light can escape from it. If somehow we can imagine living inside this concentrated incandescence, the prevalent brilliance must be of such a magnitude that it blinds us. I feel that we are in just such a predicament today, in spite of all the knowledge we possess. There is so much information around us, buffeting us from all sides, and the energy of this 'black hole' sucks in more and more knowledge, to the effect that we flail about in its black dazzle suffering from information overload. The 'black hole' of these times manifests a modern mindset, when on the one hand we build a particle accelerator known as the Large Hadron Collider (LHC) to test different theories of particle physics, whilst on the other we ignore the processes of ecosystem collapse that these very same activities generate. The LHC is an industrial-scale scientific experiment and is reputedly the largest machine in the world. It is in the form of a circular tunnel running for 27 km (16.5 miles) 100 metres below the surface of the earth. This undertaking, and future similar 'atom-smashers' may have had a colossal environmental impact,[32] but that in a sense is not the main issue at this point. The LHC's purpose is to allow:

> ... scientists to reproduce the conditions that existed within a billionth of a second after the Big Bang by colliding beams of high-energy protons or ions at colossal speeds, close to the speed of light. This was the moment, around 13.7 billion years ago, when the Universe is believed to have started with an explosion of energy and matter. During these first moments all the particles and forces that shape our Universe came into existence, defining what we now see.[33]

I do not want to take anything away from the great minds that are committed to this project, and the thinking, energy and effort that has gone into making it a reality, but how does one confront another emerging reality? What is the point in knowing what happened 13.7 billion years ago to within a billionth of a second if another group of scientists is predicting imminent ecosystems collapse (see chapter 6, 'Surviving the Anthropocene').

In a sense, this discussion has now moved forward to consider the possibility of extinction (more like species-level suicide) as a possible outcome of human conduct: 'What we do know is that, given everything ... we are living through a confluence of events that will shake the foundations of civilization, and jeopardize our capacity to sustain large populations of humans. There is enough certainty around these issues to justify being existentially alarmed.'[34] So where is all this knowledge taking us? What will happen to the PhD dissertations now being written by the thousands by all those hard-working students? (See David Orr on education below). Should clever minds not be thinking of ways of not to degrade the planet even further? There is an explanation of sorts to this storyline. We are in the age of specialisms and we burrow away in our mole-like tunnels oblivious to what is going on elsewhere. That is the nature of scientific endeavour, and the LHC mole has managed to burrow a tunnel 27 km in circumference.

We could learn from great thinkers of the past who considered themselves so overloaded with knowledge that they just had to give their brains a rest. Abū Hāmid Muhammad Ghazālī (1058–1111), known to medieval western scholarship as Algazel, was one of the greatest thinkers in the Islamic tradition. He left his post as head of a prestigious theological institute of the time and disappeared into anonymity at the height of his career. St Thomas Aquinas (1224–1274), described as one of the greatest of the medieval philosopher-theologians in Christendom and reputed to have written more than eight million words, underwent an experience in church and stopped writing during the last days of his life. What moved these great men to stop producing more light? Was there so much of it already that it was causing myopia?

Events began to take a turn in sixteenth-century Europe, which increased the dazzle considerably and which is self-described as the European Enlightenment. It supposedly inaugurated the rationalist movement, the age of reason, of thesis, antithesis and synthesis, of conjecture and refutation, of, according to Immanuel Kant (1724-1804),[35] the emergence of man from his self-imposed infancy, thus demonstrating an arrogance of monumental proportions matched only by Thomas Babington Macaulay. Kant apparently spent his entire life hardly leaving Königsberg (in present day Russia since the Second World War) his home town where he was born. His approach may have been partly a reaction to the Middle Ages (a Eurocentric description of time) also known as the dark ages, but by this sweeping remark Kant managed to brush all other traditions, including the Confucian, Zen, Daoist, the Indian Vedic and post-Vdic traditions, Buddhism and Islam into the undergrowth. Later European thinkers like Hegel (1770-1831)[36] used the tools of reason to critique rationalism itself and Nietzsche (1844-1900)[37] dismissed it almost out of hand. John Ralston Saul observes:

> Reason began, abruptly, to separate itself from and to out distance the other more or less recognised human characteristics - spirit, appetite, faith and emotion, but also intuition, will and, most important, experience. This gradual encroachment on the foreground continues today. It has reached a degree of imbalance so extreme that the mythological importance of reason obscures all else and has driven the other elements into the marginal frontiers of doubtful respectability.[38]

Nevertheless, the last five hundred or so years has managed to become known as the 'age of reason', despite the fact that past civilizations had also used this faculty in their evolution. As Max Horkheimer (1895-1973) said: 'Reason for a long period meant the activity of understanding and assimilating the eternal ideas which were to function as goals for men. Today, on the contrary, it is not only the business but the essential role of reason to find means for the goals one adopts at any time.'[39] In other words, the end justifies the means, not the other way around. This is how we have ensured the emergence of the Hitler's and Stalin's of our time.

Thinkers in sixteenth and seventeenth-century London and Paris were stretching their minds to explain the influences of their times. Francis Bacon (1561-1626),[40] recognized as the original empiricist thinker and the father of modernity, advocated knowledge based on direct observation, favoured sense experience and rejected the blind knowledge of authority. He meant by this the separation of reason from revelation and thus was a critic of the Christian scholastic tradition, in regard to which he was not found to be entirely accurate: 'It would be ... absurd to deny the profound originality ... of the 14th century scholastic scientists like Buridan or Oresme'.[41] John Locke (1632-1704), following in the empiricist mould of Bacon, is considered to be the greatest and most influential of English philosophers 'whose thought became the foundation both for classical British empiricism and for liberal democracy.'[42] Locke's thought is based on the principle that the mind remains a blank until it is sensitized by experience.

The rationalist position was postulated by René Descartes (1596–1650), whose contention was that truth comes from reasoning alone. He is considered to be the chief architect of the seventeenth-century intellectual revolution, the basis of which was scepticism and doubt. However, what is not in doubt is the doubting self itself, and from this emerged Descartes' *Cogito, ergo sum* – 'I think, therefore I am'. Richard Tarnas opines that Descartes helped emancipate the material world from its connection with religious belief, thus allowing science to evolve uncontaminated by theological dogma. The human mind and the natural world were independent of each other and also from God. The thinking mind, the subjective experience, is essentially different from the objective world of matter, although they found their common source in God (Descartes was a believer). Science was the fruit of this dualism, which included 'science's capacity for rendering certain knowledge of that world and for making man "master and possessor of nature"'.[43] This way of thinking produced spectacular tangible results; that progress appeared inevitable; that mankind's happy destiny at last seemed assured as a result of its own rational powers and concrete achievements; that the quest for human fulfilment would be propelled by increasingly sophisticated analysis and manipulation of the natural world; that mankind had at last reached an enlightened age. This is the genesis of the man and nature dichotomy. This signified a fundamental shift of character in the western psyche and 'the direction and quality of that character reflected a gradual but finally radical shift of psychological alliance from God to man'.[44]

Science to the fore

Isaac Newton (1642–1727), reputed to be the father of modern science, is credited with having brought together the inductive empirical tradition of Bacon and Locke and the deductive rationalism of Descartes.

> Newton cemented together these twin epistemologies into the experimental methodology of science: interfere with nature in someway; observe, measure and collect data; ruminate and reason a while; formulate an explanatory model or hypotheses and use it to make a prediction; finally, test the prediction and validity of the model by further interference and observation. After Newton, science became defined as experimental philosophy, the practical investigation of the organisation of reality.[45]

As Robert Hamilton observes, the crux of the Newtonian approach was to rely on the predictive power of mathematics. Quantification became the basis of the scientific order. The world was reduced to a clockwork machine which did not have a place for quality, consciousness, intuition, feeling. 'At the end of the seventeenth century, classical science stood like a steamroller at the top of a very long hill. Newton released the handbrake ... [and] the steamroller squeezed the world of quality right out of the picture.'[46] It is worthy of note that both Descartes and Newton were devout Christians. They nevertheless laid down the foundations of a scientism that eventually challenged faith and is now being held up as the alternative to religion. But all science does is attempt to explain the physical world. Faith helps us to face it.

Here then is a brief overview of the baggage that Europeans carried with them as they ruled the waves, imperiously destroying viable lifestyles that had existed for millennia, causing irrevocable destruction to huge swathes of the natural world along with the extinction of innumerable species. This is the result of black hole thinking and is the end game resulting in climate change and the destruction of the biosphere. We pride ourselves in our dedication to modernity, the terminal consumerist paradigm and the ostensible benefits it has brought us, with little regard to the natural world. We are unable to see, or perhaps we ignore, the consequences of the obvious contradictions between the endless promotion of an economic growth agenda on the one hand, and promoting a conservationist mindset on the other. Alain Touraine observes:

> In its most ambitious form, the idea of modernity was the assertion that men and women are what they do, and that there must therefore be an increasingly close connection between production, which is made more efficient by science, technology or administration, the organisation of the society governed by law, and a personal life motivated by both self-interest and the will to be free of all constraints.[47]

Self-interest free of all constraints is usually expressed as enlightened self-interest, and supposedly releases the individual from their antiquated shackles of the past to be discarded and replaced. In one sense enlightenment rationalism was the undertaker that buried the Divine alongside faith and tradition, while in another it was the midwife that helped deliver modernity and secularism.

However, the tools of rationalism were used to critique rationalism itself by philosophers such as Hegel and Nietzsche, as observed earlier, and more recently by philosophers such as Adorno (1903–1969) and Horkheimer of the Frankfurt School of critical theory, who make the bold assertion that the enlightenment project is on course to destroy itself by the very institutions it creates. This mode of thinking, and 'the social institutions with which it is interwoven, already contains the seed of the reversal universally apparent today. If enlightenment thought does not accommodate reflection on this recidivist element, then it seals its own fate.'[48] Francis Bacon's utopian vision was the human domination of nature on a vast scale. But this very vision of human sovereignty over the natural world, 'can now become the dissolution of dominance. But in the face of such a possibility, and in the service of the present age, enlightenment becomes wholesale deception of the masses.'[49] The institutions that emerge from Enlightenment thinking become the sources of the destruction of enlightenment itself. 'Horkheimer and Adorno saw technological domination of human action as the negation of the inspiring purposes of Enlightenment.'[50]

But, where were the Muslims in this discourse? Although in common with people of the other traditions they were not only profoundly impacted and broken from their moorings by modernist theology, they have also been unable to escape its vortex and now struggle to re-orientate themselves. However, some reflection would lead us to the conclusion that Muslims are not as detached as one thinks from the processes that have led us to the formulations we now describe as the Enlightenment and modernity. Thought as one of the facets, the human story has a habit of stacking up, absorbing and discarding elements as it grows. The Islamic intellectual tradition is an important layer in this process and I shall attempt to delve into this in the discussions that follow.

The Bridge

The middle nation

As has often been said we are presently living in interesting times but it also needs to be acknowledged that the times are also dangerous, not least because of climate change. There is currently an active distortion of perspective concerning relations between Islam and the West articulated by a growing fringe on the political right. Manifestations of this distortion are the emergence of an anti-Islamic party in Germany; the shutting down of six mosques in Austria on the grounds that they are foreign-funded (this ignores the existence of hundreds of foreign-funded churches in Muslim countries, and one can only hope that the authorities in these countries have sense enough not to retaliate with 'like with like'). The UK, which is regarded as more liberal than other European countries in its treatment of minorities, has experienced a rise in attacks against mosques, physical abuse against Muslims, anti-Muslim graffiti at a university, a rise in verbal abuse and hate mail, reports of Islamophobia in schools and continued online abuse against Muslims through social media.'[51] After any 'terrorist' incident in the UK and elsewhere, there is always a surge in hate crime against Muslims, but the fact remains that such activities are the acts of a minuscule minority purporting to follow an Islam counterfeited for their own purposes that is aimed at distorting Muslim relationships with the rest of the world. If anything Muslims have always considered themselves to be the middle nation: 'Let there be a community among you that calls for what is good, urges what is right and forbids what is wrong; those are the ones who have success' (3: 104). Muslims are a middle nation in the geographical sense, in that it is located between east and west, but this is also true in the social, cultural and intellectual senses.

Islam is sometimes referred to as an Abrahamic faith, in common with its sister religions Judaism and Christianity. In this sense it is a Western faith, but in the common parlance reference to the Judeo-Christian ethic reflects the two-part Bible (Old and New Testaments) Christians use. The Qur'an makes frequent references to the Torah and the Gospel, and there are recurring references to all three texts in one sentence. All the major prophets mentioned in the Bible, from Abraham to Jesus, are also mentioned in the Qur'an, but the current use of the term Judeo-Christian is usually intended to convey an exclusive European/Western ethos (Israel is considered to be part of the West although it is in the Middle East. It engages in many Eurocentric activities, including the Eurovision song contest). So Islam is of the West, and at the same time it isn't. But is it of the East? Convention and popular attitudes suggest that it is as seen through European eyes. It is of the Orient, exotic, mysterious and occasionally dangerous; hence orientalist, or orientalism. In reality Islam is both of the East and of the West, epitomized in the verse, '*The Lord of the two Easts and the Lord of the two Wests ...*' (55: 17), a geographical, cultural, social and intellectual bridge between the Orient and the Occident. The Chinese inventions of paper, printing, gunpowder, the compass and much more all arrived in Europe thanks to the Muslim land bridge. But there has also been traffic the other way: Muslims taking ideas from the Occident and carrying it in the opposite direction. In my view, Muslims have an equal claim to that of the Greeks for the evolution of Europe, and this relationship has been consistently downplayed to the detriment of all. For instance, it would be interesting to know how much of the 800-year-old Islamic presence in Spain is taught or known to the Spaniards, let alone the rest of Europe. By taking this position I also need to accept Muslim (not Islamic) culpability for the state we are in today. And for now I would just like to deal, in the briefest of detail, with a sliver of the Islamic intellectual tradition that has profoundly influenced Europe in its lead up to the Enlightenment.

Enlightenment philosophers drew a line between reason and revelation. They considered revelation to be incompatible and a barrier to the evolution of thought and consequently to civilization itself. This was when the Divine was relegated backstage in the affairs of the human race. Debates and disputes on the relative merits of reason and revelation existed from the earliest times in Islamic intellectual history. The Mu'tazilites, founded in the second century after the Prophet Muhammad's death, were arch Islamic rationalists and they debated endlessly about free will (*qadar*) and predestination (*jabr*). The Ash'ari and the Maturidi doctrines,[52] which are now part of the Islamic mainstream, were a reaction to the excessive rationalism of the Mu'tazilites who are now regarded as heretical by Sunnis, though their influence continues amongst the Shiites.[53] As Arberry argues, the Islamic position was to treat reason and revelation on their own terms and they were complementary. In Islam, reason may be viewed as encouraging intellectual freedom while revelation made space for contemplating the Divine order. Muslims in search of first principles always hark back to revelation, the Qur'an, and thus: '*He (Allah) created man and taught him clear expression*' (55: 2, 3); the Arabic term *bayān* is variously translated as 'expression', 'intelligence', 'communication', 'speech', or 'explanation'. Or, *'Surely in the creation of the heavens and earth and in the alteration of night and day there are signs for men possessed of minds'* (3: 190). This is the Qur'an explaining itself and thus, using the mind, reason is accepted 'as an ally of faith'.[54] The tripartite division of knowledge into the rational, empirical and the revealed, although not expressed in these terms, can be seen as being inherent to the Islamic processes of enquiry as evidenced by the evolution of scientific enquiry (discussed below) in the Islamic world.

Al-Kindi (d. after 870) was the first known Islamic philosopher and a leading proponent of the Mu'tazilite school of Islamic theology. He was instrumental in translating Greek philosophical works then lost to Europe and examined the 'distinction between human and revealed knowledge'.[55] By this time Christian scholars were already thinking about the relationship between reason, revelation and the wider theology that was beginning to occupy the minds of the emerging Islamic philosophers. Al-Farabi (Alfarabius, 870–950) is described as an 'Islamic Neo-platonist philosopher of language, culture and society'[56] and wrote commentaries on the works of Aristotle. There was a reaction to this trend by the orthodox Sunni theologian al-Ash'ari (873/4–935/6). Starting off as a Mu'tazilite himself he reacted to 'the onslaught of Hellenistic thought ... [and he exploited] Greek dialectic for his own orthodox ends'.[57] The philosophical tradition was given a further boost with the emergence of Ibn Sina (Avicenna, 980–1037). Described as one of the greatest thinkers ever to write in Arabic, his works were 'cited throughout most mediaeval Latin philosophical and medical texts.'[58] This was met by a further challenge from orthodoxy in the person of al-Ghazali (1058–1111) who 'furthered the triumph of revelation over reason'[59] and attacked Ibn Sina in his *The Incoherence of the Philosophers* (*Tahāfut al-Falāsifah*). An Ash'arite, he fought fire with fire and called on theologians to use the philosophical method to refute heretical arguments. The philosophers were not about to give up. Ibn Rushd (Averroes, 1126–1198), who was not part of the Mu'tazilite theological school, was an acclaimed Aristotelian commentator and had an 'enormous influence ... in medieval Europe, which was rocked by a wave of "Averroism" when 15 of his 38 commentaries were translated from Arabic into Latin in the 13th century'.[60] His works were studied in the West to the mid-seventeenth century. He was an Islamic judge (*qādī*) during the period of high culture in Islamic Spain and rebutted al-Ghazali in his *Incoherence of the Incoherence* (*Tahāfut al-Tahāfut*).

It could be argued that the three major tectonic plates of Islamic thought had defined themselves by the approach of the fourteenth century. The first was Mu'tazilite, a school of Islamic theology (*kalām*) that had spent itself out by this time but continued to wield influence outside the boundaries of Sunni orthodoxy. The second were the Sunni schools, represented by the Ash'ari Maturidi orthodoxy. Ash'ari was once a Mu'tazilite himself and he employed its techniques in his own work, which was severely critical of the positions the Mu'tazilites were taking. The Ash'ari and Maturidi (853-954) return to basics was further shored up by Ibn Taymiyya (1263-1328), who displayed 'in his polemical broadsides a superb mastery of the methods of dialectical reasoning'.[61] The third tectonic plate is Islamic mysticism, or Sufism, as it is popularly known, which had developed a teaching method that led to a personal experience of God. It even received the approval of Ibn Taymiyya, who was reputedly a member of the Qadiriyya Sufi order founded by the twelfth century mystic, Shaykh Abd al-Qadir al-Jilani (1078-1166).[62] Which raises an interesting question: was Ibn Taymiyya the arid literalist he is made out to be?

In these exchanges, men of light had been scurrying to and fro over the bridge between the Orient and the Occident, and by the end of the twelfth century the Latins had translated as much as they wanted or cared to from the Arabic. Although the lights of Muslim intellectual discourse didn't entirely die out they were beginning to dim after this episode of Islamic history. The reasons put forward for this diminution was the emergence of a literalism purported to derive from the teachings of Ibn Taymiyya and typified today by the neo-literalist Salafi/Wahabi movements, a retreat from the exercise of personal judgement by jurists on legal matters, sometimes misleadingly referred to as the closing of the gates of *ijtihad*,[63] and the sacking of Baghdad by the Mongols. There is much scholarly exchange today about the causes of this decline, some arguing that there has been no decline, but they remain an esoteric conversation far removed from the Muslim mainstream.

By the thirteenth century this spirit of enquiry had moved on to Europe and the translated works of Muslim scholars were being avidly studied by their Christian counterparts in Europe:

> The Rendering into Latin of the Arabic translations of Aristotle, and, much more importantly, of the Arabic commentaries on his work by the major Islamic philosophers ... provided the catalyst for an explosion of Aristotelianism in western Europe ... Thomas Aquinas (1225-74) tried to produce an acceptable synthesis of Christian thought and Aristotelianism, having become acquainted with the thought of Aristotle and Averroes [Ibn Rushd] while a student in Naples.[64]

Ibn Taymiyya was just eleven when Thomas Aquinas died and there is no evidence to suggest that the former was aware of the latter's enthusiasm for the self-same Ibn Rushd he had criticized and whose thoughts 'would rise, phoenix-like, in the West within a few years to disturb, perplex, and challenge another orthodoxy'.[65] But there were also the likes of Ibn Taymiyya in the Christian scholastic world of the time who saw Thomas Aquinas' work as tainting Christian orthodoxy. Nevertheless, his work was accepted in time and he was canonized in 1323. The thread I am following takes me from the Islamic Mu'tazilite tradition through to Christian scholasticism to Renaissance humanism and finally to the Jesuit scholars, under whose tutelage the arch-rationalist René Descartes was nurtured. Ibn Sina and Ibn Rushd were still being studied in European institutions at the time, and thus the Islamic philosophical tradition of Mu'tazilite scholars and their successors played no mean part in the advent of the Enlightenment discourse which is seen to have ushered in modernity.

The sacred and the secular

The investigation of natural phenomena was an ancient occupation known to the Chinese, Egyptian, Greek, Indian, Mayan and other cultures. It was the inevitable outcome of innate human curiosity. This area of knowledge was known as natural philosophy, before the word 'science' became fashionable in the early nineteenth century. Astronomy may have started as an exercise in curiosity, but travellers have known how to set their course by the stars for millennia. Similarly, medicine evolved from simple folk cures to become a major scientific pharmaceutical endeavour and hydrology from the need to understand simple irrigation systems by ancient agriculturists to the complex specialism it is today. Many other specialisms emerged as civilizations evolved, and the forces that drove them had nothing to do with economic or industrial progress.

The great Muslim advances in astronomy had much to do with discovering the times and directions of prayer. Islamic natural philosophers (scientists) who bridged the gap between the ancient and the modern took advantage of existing knowledge and developed methods of their own. As Roberts observes, 'To no other civilization did Europe owe so much in the middle ages as to Islam'.[66] Muslim scientists were developing experimental techniques unknown before their time and the terminology used by them pre-echoed the twin epistemologies of empiricism and rationalism. *Istidlal* implied experimentation, measurement and observation; *istiqrar* is identical to the empirical, inductive method which was used 500 years before Bacon; *istinbal* could be described as the analytical method.[67] Ibn Sina's textbook on medicine was a standard work in Europe until the nineteenth century. Al-Haytham (Alhazen, 965–1040), who is now considered to be amongst the earliest theoretical physicists, founded the science of optics. Would the Hubble orbiting space telescope that is revealing much of the secrets of the universe today be available to us without his original work? In the twelfth century, al-Khazini recorded the specific gravities of fifty substances, which compare remarkably well with today's results.[68] There was no interest in this subject in Europe until Robert Boyle conducted his experiments 500 years later.

The Qur'an added to the motivation of finding the direction and times of the daily prayer by encouraging the discovery of the secrets of the universe. It was knowledge sought for knowing creation and acknowledging the Divine: '*He (Allah) created everything and determined them in exact proportions*' (25: 2). *Taqdiran* is variously translated from the Arabic as 'very exactly', 'in due proportion', or 'exact measure', so revelation can be seen as encouraging the search for exactness in knowledge and the development of the critical factor innate in the human; a scientific process, as we might understand it today. Would it have been possible for Copernicus to discover that the Earth moved round the sun and overturn Ptolemy's geocentric system, which had been accepted for thirteen centuries, without the intervening contributions made by a succession of Islamic astronomers and mathematicians? 'The way was paved by the gradual chipping away at the edifice of the Ptolemaic system over the centuries by countless Arab [Islamic] astronomers, both with their observations and their often ingenious theories.'[69]

Would it have been possible for Descartes to formulate his mathematics without the work of Mohammed ibn-Musa al-Khwarizmi, for example? In the ninth century, al-Khwarizmi was the first mathematician to work on equations that equaled zero, or algebra (*al-jabr*) as it has come to be known. He also developed quick methods for multiplying and dividing numbers, known as algorithms (a corruption of his name). Al-Khwarizmi called zero '*sifr*', from which cipher is derived. By 879 CE zero was written almost as we now know it, an oval, but in this case smaller than the other numbers. 'And thanks to the conquest of Spain by the Moors, zero finally reached Europe; by the middle of the twelfth century, translations of al-Khwarizmi's work had weaved their way to England.'[70]

Today we are swamped by algorithms, from the type of advertisements we see on social media applications in the electronic devices we use to Wall Street trading and the search for correlations in the data to identify crime 'hot spots' by police forces. The zero crossed the bridge from India and Muslim mathematicians made today's calculations possible by further exploring its possibilities and introducing the decimal point to take care of fractions. Both these mathematical innovations then crossed to Europe by the twelfth century. However, as Lynn White Jr states, 'By the late 13th century Europe had seized global scientific leadership from the faltering hands of Islam.'[71]

But whither the natural world in all this? Or environmentalism? Traditional and indigenous communities considered themselves to be integral to nature, and in the absence of scripture the natural world itself was the text. The Eastern traditions also had a close affinity with nature and it was looked upon as a gift from the Creator in the Abrahamic faiths. All these ways of being and relating to the natural world were destroyed when modernist thinking emerged and turned it into an economic resource. Nature is an entity in whose embrace we exist, as exemplified by the Sufic tradition in Islam and by the Franciscan tradition in the Christian faith. So what was it that led Lynn White Jr to assert that, 'Christianity bears a huge burden of guilt'[72] for setting in train this unprecedented assault on the natural world we are now witnessing? Was he comparing the Cartesian position of being the 'master and possessor of nature' with the Christian biblical, 'fill the earth and subdue it'?[73] After all, Descartes was a believer. Seyyed Hossein

Nasr observes: 'neither Christian Armenia nor Ethiopia nor even Christian Eastern Europe gave rise to the science and technology which in the hands of secular man has led to the devastation of the globe'.[74] So this is a particularly Western European affair and White's observation bears some weight:

> Modern science is an extrapolation of natural theology and ... modern technology is at least partly to be explained as an Occidental, voluntarist realization of the Christian dogma of man's transcendence of, and rightful mastery over, nature. But ... somewhat over a century ago science and technology – hitherto quite separate activities – joined to give mankind powers which, to judge by many of the ecologic effects, are out of control.[75]

The causes of the all-out war we are now waging against the environment can be reduced to just two main causes, from which all other causes that gave us modernity have emerged. The first of these, as we have seen earlier, is the loss of a sense of the Divine and the sacred, which kept us anchored within the limits of the natural world. We now exist in soulless secularized spaces, concealing the reality that we are all trapped in an irresistible undertow of debt and hedonism aimlessly driving us through oceans of consumerism. The philosopher John Gray opines, 'Secular societies are ruled by repressed religion. Screened off from conscious awareness, the religious impulse has mutated, returning as the fantasy of salvation through politics, or – now that faith in politics is decidedly shaky – through a cult of science and technology.'[76]

The second cause, which derives from the first, is the human disconnection from the natural world that occurred from the moment we perceived it as 'other'; an exploitable resource which is now expressed as 'ecosystem services'. As George Monbiot observes in an opinion piece published in the *Guardian*:

> The natural capital agenda is the definitive expression of our disengagement from the living world. First we lose our wildlife and the natural wonders. Then we lose our connections with what remains of life on Earth. Then we lose the words that described what we once knew. Then we call it capital and give it a price. This approach is morally wrong, intellectually vacuous, emotionally alienating and self-defeating.[77]

Enlightenment thinking conjured nature into another space, separated it from the human and objectified it. We are now alienated from that which sustains us. There was no longer anything we could do to nature that would invite censure and all we needed was a device that would help us get the most out of it in the shortest possible time. This device was money, more precisely money as we understand it today; the product of a clever illusion, the cement that holds our fragile house of cards together. This is difficult to digest given what it means to all of us in our daily lives, but it does require some examination given its all-pervasive nature. Starting this conversation may even persuade us to look at solutions that go to the root of the matter. As a medium of exchange, the money we use today is of a special kind. It is essentially 'non-money', created historically by what has been defined by economists as the fractional reserve banking system, which has been further corrupted in more recent times and rules our lives today from cradle to grave. Money has now replaced the space once occupied by the Divine. We worship it. Money is the universal God of our times, it conjures up for us the possibility of heaven on earth, it seeps into every corner of our lives and its virus-like nature is devouring its way through the natural world, thus leaving behind a degraded Earth for future generations. (Because of its all-pervasive nature I discuss money in various parts of this book. See particularly the last section of this chapter and the section under 'Acceleration' in chapter 4.).

Was there no environmental destruction prior to the advent of modernity, the Industrial Revolution, the European voyages? Of course there was and it began the moment a human saw the potential in poking a hole in the ground, placing a seed in it and watching it grow. This must have been the moment it dawned on the human species that the environment can be manipulated. As civilizations waxed and waned they ravaged the environment around them for their own ends and when they ultimately died out the natural world renewed and came alive again. Profligacy was contained by the twin boundaries drawn by acknowledging a power greater than that of the human, expressed diversely throughout myriad societies, along with a deeply engrained empathy with the natural world and a mode of being that responded to its rhythms. Planet Earth was reasonably safe until the human species discovered a method of conjuring money out of thin air.

There is another element, which is as vital to our present day lifestyles as the form of money we use and is closely related to it, and that is energy. To lead an ordinary life today there are just a few things that are left for us to do without flicking a switch, or pushing a button, or turning a knob. Our addiction to fossil fuels has made life comfortable for us but it has led to the warming of the planet, with all its attendant consequences, which we are only now trying to come to terms with. The nuclear option appears to be one way out of the fossil fuel dilemma, but this is leading to the creation of other hells for the generations that succeed us. Two examples of this are the Chernobyl[78] and Fukushima[79] nuclear disasters. These places were the locations of nuclear power generators, the former built by the once formidable pseudo-capitalist Soviet Union and the latter by Japan, to meet the insatiable energy demands from industry and consumers to keep modernity on the move. Chernobyl is now a no-go area for human habitation, as large spaces surrounding the concrete sarcophagus enclosing the damaged reactor are contaminated by dangerous radioactivity that will persist for a few thousand years. Nature has largely reclaimed the area, but sadly the long term effects of the radiation on other life forms are yet to be seen and people are prohibited from entering the exclusion zone. As for Fukushima, it too has its no-go areas and additionally, in this case, radiation leaks are at a level that are clearly capable of contaminating the entire Pacific Ocean and depriving a few million people of their livelihoods and source of protein. The health hazards that emerge from this don't bear contemplation.

We may manage to reverse climate change, providing we follow the international commitments that have been agreed upon – American spoiling tactics notwithstanding – but how are we to escape the radiation from the nuclear waste that we are creating that will last for thousands of years? Also, just to remind ourselves, renewable energy can only save us if we free ourselves from our addiction to consumerism and stop chasing after economic growth on a finite planet, which in its very essence is anti-conservationist (see chapter 3, 'Prosperity in Perspective').

The environment as an '-ism' was not a topic that seriously occupied people's minds until the second half of the twentieth century. This is not the same as saying that there was no interest in nature or the love of it in the past, and this would be an absurd position to take, but we had to wait until then to realize that something was awry. The planet's carrying capacity absorbed the body blows it was subjected to by European global expansion from the closing years of the fifteenth century, by its assault on the natural world by the nineteenth century capitalist Industrial Revolution, by the twentieth century communist (pseudo-capitalist) industrialization, by the two great world wars and by post-war progress (see chapter 3, 'Prosperity in Perspective') as defined by economic growth. Then people began to notice, in the last decades of the last century, that we had been taxing the planet beyond what it can endure. It is generally accepted that Rachel Carson's seminal *Silent Spring*, published in 1962 (see chapter 2, 'A Delayed Reaction'), was the consciousness-raising trigger and since then we have been trying to get to grips with what we have brought upon ourselves. Einstein was supposed to have said that the thinking that caused a problem cannot be expected to be the source of its solution, or words to that effect, a perfectly rational observation to make and I have heard others speak in similar terms in the past. But it would seem that a congenital aversion to common sense does not allow us to learn from past experience, though Einstein's name does give some weight to the proposition. As it is often said, we are living through the most brilliant civilization the human race has managed to create. But in spite of all the even more brilliant advances being conjured to take it to further heights, a consensus is now emerging that our civilization is now in terminal decline. Discovering what made it great is also to uncover the causes of its demise.

A Lethal Cocktail

Makings of a monoculture

The European voyages we have been discussing were part of an even bigger picture. At the same time the Europeans were scouring the seas to plant their flags in lands that were ostensibly empty, much else was happening back home. I leave an explanation of the causes of the seminal events in Europe that led to European global hegemony to learned historians, but the fact remains that the Europeans were pushing out in many directions, to some of which I have given cursory examination. Europe was coming to terms with itself at the same time as it was consolidating an empire, and the evolving European experience in terms of the cultural, social, intellectual, political, technological and economic was imposed upon the rest of the world regardless of whether or not this experience was germane to local conditions. This stunted the flowering of local societies and their cultures in their own way and in their own time. The loss of cultural diversity preceded the loss of biodiversity by centuries. Cultural diversity was reduced to a slogan and the world is now poorer for all those rich expressions of life we have lost. As Europe changed it was changing the rest of the world with it, and the raison d'être for this mindset was the divinely inspired civilizing mission, bringing hope and progress to benighted natives – without of course losing sight of the twin motivators of power and profit:

> The result was always uncomfortable and often brutal. It is also true that religious zeal could blur easily into less noble motives. As the greatest Spanish historian of the American conquests put it when describing why he and his colleagues had gone to the Indies, they thought 'to serve God and his Majesty, to give light to those who sat in darkness and grow rich as all men desire to do'.[80]

The paradox is that the mindset of the civilizing mission also carried within it the germ of its own destruction.

European imperialism had a logic and method all its own: A bridgehead was established in a 'newly discovered' land; it was then explored for plunder; territory was consolidated militarily; new diseases brought in by the Europeans took hold and wiped out large chunks of the indigenous population; the introduction of alcohol debilitated the survivors; agents took over and established political control; entrepreneurs and commercial interests established trading posts and monopolized trade; surveyors mapped the land as a preliminary to resource exploitation; prospectors and miners went after the minerals; loggers felled forests; settlers arrived and planted commercial crops such as tea, rubber, tobacco, sugar cane and cotton. Somewhere along the line missionaries stepped in and established schools giving life to the dictum, 'Give me a child until he is seven and I will show you the man' (this statement is usually attributed to St Francis Xavier founder of the Jesuit missionary order). Inherent in this process of Christianization was the westernization and secularization of cultures resulting in the extinction of the traditional and the indigenous. Sometimes this was achieved by the forcible abduction of children, who were virtually incarcerated in mission stations far from their parents and communities. Thus in a generation the whole ethos of a culture and its knowledge base was destroyed and then rebuilt in a form alien to it. The human ecology of the region was ripped apart along with its natural habitat. It was a prolonged dying out.

Those who rebelled were branded as savages and suffered the consequences, or chose to live in reservations – the human equivalent of wildlife parks – reviving memories of an irredeemable past from generation to generation. Others were forced into slavery, and yet others transported as indentured labourers to build railways and roads and work in plantations. The bulk of those who remained became hewers of wood and carriers of water, as wage slaves, and yet others became part of the native soldiery and served the purposes of their colonial masters. A minority became part of a cadre of native administrators and technicians to run the middle and lower rungs of empire and, having lived as subjects both physically and psychologically, the old ruling elite raised their aspirations to occupy higher positions in the colonial hierarchy. Thomas Babington Macaulay must be grinning in his grave, for his interest was not so much the Christianization of natives but rather the production of willing colonial subjects that would help run Empire and the mission schools were part of the process.

European imperialism was in the nature of a social engineering project on a grand global scale, built on the graveyard of existing civilizations and cultures. Neither was there any escape from this during the decolonization process, as by this time the alien European outlook had so saturated the psyche of the old colonials that they identified with it without much equivocation thus fulfilling the hopes of their masters. Thus the progress of our present civilization in the past 500 years has been unidirectional in the sense that the human race has been acting as one in a lethal alliance in its assault against the natural world.

The established major traditions survive today only in their private ritualized forms. For example, is it possible today for the full traditional forms of Buddhist, or Hindu, or Shinto, or Taoist or Islamic economics, lifestyles and cultures to survive let alone thrive in the present global order? The global system now in place makes it impossible for this to happen as the entire Earth's surface is covered with an invisible, impregnable secular layer resembling concrete. Faith has been reduced to an enactment of religious rites based on moon sightings and the annual calendar. At the level of the profane, culture has become a nostalgic expression of the past on this impervious concrete stage. Now part of the entertainment industry it peddles nostalgia for profit – cultural festivals, world music, fusion food and so on. At the international level, Faith Based Organizations (FBOs) are at long last permitted to participate, for example in climate change events, at the discretion of secular bureaucrats. The Alliance of Religions and Conservation,[81] for example, is an avowedly secular organization. Sometimes cracks do occur in the concrete and green shoots reach upwards in the sunlight like the Amish, 'a group of traditionalist Christian church fellowships'[82] in Pennsylvania, USA, but they too often emerge from the cracks and inevitably walk over the concrete as ants seem to do in search of forage. The cracks will widen as the current system inevitably collapses, and this may provide spaces for some traditional cultures to re-emerge if only the opportunity is grasped. This will also provide scope for alternative models to spring up, and some are already doing so. Though the so-called 'Islamic State' group that emerged in 2014 in the vacuum created by wars in Syria and Iraq has excluded itself from any consideration as an alternative model for Muslims by its penchant for unbridled savagery. It resembles a degraded bush of thorns without roses.

Be that as it may, this cocktail consists of an educational system that is fundamentally consumerist in its ethos, which in turn leads people along career paths intended to create an obliging professional class that consume their lives away; a political model that would have been appropriate for Europe in its evolution but not for other civilizations with different traditions; a financial system that is questionable; and a commercial system that is fundamentally exploitative. Collectively, they have led us – just in the past 500 years – to batter the Earth in a manner previously unknown since Homo sapiens began to till the land more than 10,000 years ago.

Traditional education, which was mainly concerned about giving children a sense of place, has been replaced by a universal secular model likened to a factory system that matched its industrial counterpart. Today, the education conveyor belt supplies people graded through a quality control system determined by examinations for the producer/consumer capitalist global model. As the twentieth century matured, this system so entrenched itself to the extent that we now have international schools and branches of European and American universities established in most so-called developing countries turning out an end-product who seeks prosperity, progress and development. This is a far cry from the missionary schools of old. A system that once produced Christianized, obedient colonial servants is now one that produces obliging, secularized consumers.

Education as we understand it today is seen as a panacea, but is in fact a double-edged sword. Its aspirations are to give people a good life, but today we live in an increasingly overcrowded planet. There are many more of us demanding an exaggerated view of the good life, which would have been unthinkable in the times of our forebears, thus taxing the earth's ecosystems more than it can endure. As the educationist David Orr observes:

> The truth is that many things on which your future health and prosperity depend are in dire jeopardy: climate stability, the resilience and productivity of natural systems, the beauty of the natural world, and biological diversity.
>
> It is worth noting that this is not the work of ignorant people. It is, rather, largely the result of work by people with BAs, BSs, LLBs, MBAs, and PhDs.[83]

Orr observes that 'Francis Bacon's proposed union between knowledge and power foreshadows the contemporary alliance between government, business, and knowledge that has wrought so much anxiety and mischief'[84] and points a finger at the Enlightenment asserting that it 'laid the foundations for modern education, foundations now enshrined in myths we have come to accept without question'.[85] To Orr's tripartite union between government, business and knowledge I would add one other vital element, and that is the banking system. I will return to this after some reflections concerning the nature of democracy.

The colonies revolt

In the early days of empire it was the monarchies that ruled the roost and early settlers ran their new possessions by monarchial writ. A substantial amount of plundered wealth poured into the coffers of the monarchs, and in addition to this settler administrations were expected to pay taxes to the rulers of their respective parent countries. This was challenged in the North Atlantic theatre of colonial expansion. The culmination of the struggle by the British American colonies against taxation imposed by the British Parliament was their Declaration of Independence in 1776. Historians identify the American Revolution as the movement that gave democracy new meaning and 'the adoption of the democratic principle in 1787 was immensely important and justifies the consideration of the constitution as a landmark in world history'.[86] The opening words of the American constitution reads, 'We the people'. The familiar aphorism, 'set a thief to catch a thief', fits well into this narrative as the only people who were able to challenge European imperial power were the Europeans themselves, on this occasion in the guise of settlers in a distant land. The rest of us had to wait until the middle of the twentieth century to free ourselves.

The French, who were also adventuring in North America at the same time, played a key role in decoupling the British from their American colonies. It was in Paris in 1783 that the British conceded sovereignty to the burgeoning American Republic. But then the French were themselves experiencing unprecedented changes on their home ground. The first French Revolution lasted from 1787 to 1799, when Napoleon seized political power in a coup d'état and crowned himself emperor in 1804. At the end of the chaos, what emerged was a secular democratic republic based on Enlightenment principles reflected in the popular phrase 'liberty, equality, fraternity', which of course the colonized people never experienced. I refer to these events to underscore the fact that sea changes were taking place at the heart of Europe at the same time Europeans were consolidating empire. Two other events are germane to this narrative.

The Treaty of Westphalia, concluded in 1648, put an end to the religious wars of Europe; the Thirty Years' War in the Holy Roman Empire and the eighty-year war between the Spanish and the Dutch. The idea of the sovereign state emerged from this treaty. 'As European influence aggressively spread across the globe, the Westphalian principles of sovereignty and inter-state commerce became central to international law and to the prevailing world order'.[87] This was the beginning of the formation of sovereign states in Europe, which also led to the idea of sovereign territory and its subject people (I was told I was a 'British subject' as I was growing up in Ceylon, now Sri Lanka) as the imperial powers began to draw arbitrary lines on the map. Although warring continued to be a means by which Europeans (and the rest) continued to settle some of their affairs, it could be said that the emergence of nation states in Europe was organic, as we see how the French, Italian and German states evolved. They did it for themselves, in spite of the means deployed.

The rest of the world was done for by the Europeans and with the possible exception of Ethiopia the globe was carved up into artificial states, arbitrarily put together by squabbling colonial powers. In some cases they were complex conglomerates, like Indonesia, India and Nigeria, and in others curiosities like Gambia and Senegal in Africa, and Dutch, French and British Guyana in South America. In this carving up very little attention was paid to tribal affinity, linguistic commonality and religious affiliation. It was based on who ruled what, where and when: British Africa, French Africa, the Belgian Congo. Given the human disposition for conflict over territory, this has compounded the situation and demonstrated to be a recipe for permanent internecine warfare. The second event that merits mention in this context was the so called 1688 Glorious Revolution in England. It marked the tentative beginnings of parliamentary democracy when sovereignty passed from the monarch to the English parliament 'and began to function as a constitutional state'.[88]

As the art of democratic governance evolved we have had parliamentary democracy, constitutional monarchy, federations and confederations, republics, the presidential system with or without executive power. What they all had in common was an evolving capitalist system which, *a priori*, depended entirely on a banking structure that created money. Much of this took place with the vast majority of the subject people, except a tiny elite, oblivious to the changes that were taking place in the western theatre and which was being transferred piecemeal to the colonies. What the common people did experience was the monetization of their transactions, which overtook existing traditional modes of exchange. The political movements that emerged out of Marxist thinking, such as the rule of the proletariat, one party rule, socialism, communism – and their polar opposites the fascist dictatorships – would not have been able to function without state monopoly banking. My experience growing up in Ceylon (now Sri Lanka) in the 1940s was under a benign colonial administration learning the mysteries of parliamentary democracy, capitalism versus communism; jostling in debating societies with liberals, socialists, nationalists, Marxist-Leninists, Bolsheviks, Trotskyites, anarchists and the like: All western exports, like the strange spectacle of communist capitalism in China.

In the end it all turns out to be one system, the object being to suck the Earth of its resources one way or another depending on one's political affiliations to satisfy human greed: The facilitators were the banks.

Democracy and the nature of money

When decolonization finally emerged in the aftermath of the Second World War, the freedom won for the people by erstwhile freedom fighters proved to be illusory. One seminal event ensured that all the new born nation states were well and truly fettered to the international system that the West had worked so hard in creating. Alongside the legacy of parliamentary democracy, departing colonial powers felt compelled to leave behind central banks, seen as an essential ingredient of the nation state model. This umbilical cord between parliamentary democracy and the banks is of such importance that if severed would result in the collapse of the whole system. And this prompted the thinker and spiritual guide Shaykh Abdalqadir as-Sufi to observe that 'democracy was the service industry of the banks'.[89] This insight was confirmed many years later when, in the aftermath of the global financial meltdown in 2008, statements such as 'the banks are too big to fail' became common parlance in the media. What this meant was that if the banks failed the global system would collapse, and it very nearly did. It came as no surprise, then, to hear more recently that 'Washington was in thrall to Wall Street' in a popular public affairs programme on television.[90]

This has led me to another layer in the house of cards that bears some examination: the function of money in the area that is loosely referred to as market forces. Jeremy Seabrook observes that markets are 'universal mechanisms for answering human need'.[91] And he further reflects:

> When we look at the energy and effort absorbed by buying and selling in the West, at the swiftness of the transactions, the accelerating circulation of money, the besieging of the shopping malls and *gallerias*, it is clear that something is occurring which goes far beyond a mere functional answering of need. We are in the presence of something more profound and significant.[92]

Although Seabrook refers specifically to the West, we can today – without much contradiction – stretch his remarks to encompass the rest of the world too. Consumerism 'has become inextricably bound up with the roots of human identity',[93] and 'What was an important aspect of all cultures has become the universalised focus of world culture'.[94] The market economy has been 'sanctified as it has never been before' and then, 'Nowhere is the faith of the cult more ardently expressed than in that which unlocks access to limitless abundance and emancipation – money'.[95]

Here Seabrook is eloquently articulating the near godlike status we have attributed to money. But we have to refer back to Adam Smith and his seminal

Wealth of Nations to see where he has taken a great deal of trouble to convince us that money is a commodity like any other.[96] But this is the only commodity above all else that can be conjured out of thin air by the banks. Everything else, from a pin to a jumbo jet, has to be manufactured. The technical paraphernalia needed to bring money into existence and commodify it notwithstanding, it would seem that the nature of money is undergoing a transformation, according to the view of certain economists who describe it as no more than 'a social relationship of credit and debt'.[97] Money has now been elevated to the status of the air we breathe and the water we drink, but the fact remains 'that the general public is simply not aware that banks do indeed create the money supply [out of thin air]'.[98] Considering this in the light of climate change and the environmental debacle we have wrought, the money we use today transforms itself into a virus eating into the fabric of the natural world.

In the recent past this fictional money was created by ledger entries, but today it can appear as the result of a gentle tap on a computer keyboard by privileged people sitting in plush office suites: 'Those with the power to create new money have enormous power – they can create wealth simply by typing figures into a computer and they decide who can use it and for which purpose'.[99] Those of us who are familiar with electronic banking are a little closer to understanding this mechanism. Technology it seems has outstripped the fractional reserve system and has taken us further into fairytale land; some would argue Frankenstein land. Banking deregulation in the latter half of the last century 'neglected their (the banks) crucial function as creators of the money supply'.[100] Technically the fractional reserve system had its limits, but deregulation had bankers straining at the leash and what emerged was an aptly named 'balloon' of money created by the banks wound around a core of central bank money. In 2006, just prior to the financial crisis, the imbalance between commercial and Central Bank money was such that it 'fuelled much of the unsustainable credit boom'.[101]

The supply of money can also be managed by printing money, euphemistically described as quantitative easing (QE), which was found to be necessary by the North Atlantic nations to prevent their collapse in the aftermath of the 2008 financial meltdown. The natural market that Adam Smith went to great pains to define has not existed since the banking system, as we know it today, encroached on the lives of people. It has been subjected to the vagaries of economic and fiscal policies of individual nation states, of which the more powerful are expert at manipulating to their own advantage. If, as modern economists have come to conclude, money is nothing more than a social relationship then these relationships have been manipulated to the extent that the planet is now on the edge of disaster.

The psychologist Dorothy Rowe asserts that money is an idea which is dependent on the meanings we attach to it, but 'the effectiveness of our own meanings is limited by the knowledge which informs our meanings and by the meanings which other people create.'[102] For Adam Smith money was a commodity, modern economists define it as a social relationship, or merely information that manifests itself as algorithms on computer screens: This is advance warning of the emerging cashless society, and look out for Blockchain and Bitcoin. The fact that it is an entity that enables the functioning of society today means that it is brought about by a technical process that gives it 'life'. But the fact also remains that this 'life' is intangible, ephemeral. Money has also been described as a cancer. It grows inexorably until it kills its host. The idea of money that I use in this book is that of a virus – one that eats into the fabric of the natural world every time we use it. This is how we are degrading the planet.

Arguably, one possible advantage of the present system is that it makes the meeting of tax obligations to the state that much less painful in that one can write a cheque, or as it is now becoming increasingly common, to pay electronically. No more are the days of labour that have to be pledged to the monarch, or the bushels of wheat that have to be surrendered to his soldiery as signs of a citizen's loyalty.

One fascinating fact about the money we use today is that if we – from the individual to the state and every business, enterprise and institution in between – were to settle all our debts, money would just disappear. This is highly unlikely, as we are collectively not about to pay off our debt since debt is now the blood supply of the global anatomy. We have thus placed the banking system in an unassailable position. Mervyn King, one time Governor of the Bank of England, observes:

> In good times, banks took the benefits for their employees and shareholders, while in bad times the taxpayer bore the costs. For the banks, it was a as case of heads I win, tails you – the taxpayer – lose. Greater risk begets greater size, greater importance to the functioning of the economy, higher implicit public subsidies, and yet larger incentives to take risks ... All banks, and large ones in particular, benefited from an implicit taxpayer guarantee, enabling them to borrow cheaply to finance their lending.[103]

Trade and exchange based on gold and silver had survived for millennia. 'The Egyptians were casting gold bars as money as early as 4000 BC, each bar stamped with the name of the Pharaoh Menes'.[104] Gold bars were for large transactions and international trade, and even if the common people were unable to use these gold bars they represented a unit of value against which they could conduct their exchanges using token coinage or barter. Transactions were local and organic.[105] Modern day central banks ensured the monetization of the entire planet and tethered every nation and their populations to the international system. Global hegemony in another dimension was thus ensured without a shot being fired, and when the West went into financial meltdown in 2008 the whole world suffered.

Debt also laid down the foundations upon which the phenomena we have come to know today as neocolonialism has been built and globalization sustained. The twentieth century emerged with the Gold Standard intact, but the brutal First World War imposed long-term costs on the leading economies of the time. After a period of trial and error, Great Britain finally came off the Gold Standard in 1931, in the midst of the great depression that was to last till 1939 when the Second World War broke out. The war revived the economies, but what of its aftermath? The period after the First World War was bad enough, but how can global economic stability be re-established after another war just as disastrous? How can western hegemony established over the rest of the world in the preceding centuries be maintained? How to establish a permanent world peace based on trade? Out of these concerns emerged the consensus named after the small town of Bretton Woods in New Hampshire, USA, where an agreement was concluded in July 1944 at a gathering of forty-four nations. This gathering established the International Monetary Fund (IMF) and the International Bank of Reconstruction and Development, now part of the World Bank.

There are two positions on the Bretton Woods Agreement: The first encouraged an open international free market system in a politically neutral market and the second represented, 'the aggressive desire of the capitalist market to expand globally beyond the boundaries of the developed industrial world'.[106] Crucially, the Bretton Woods system established the American dollar as the currency to which all other currencies would relate, by virtue of the fact that the US had at that time the world's largest reserves of gold. Thus the Gold Standard was perpetuated by default when the forty-four nations attending Bretton Woods placed their trust in the hands of the US. The dollar was pegged at $35 an ounce of gold. The Soviet Union dismissed the whole process as a capitalist trick engineered by Wall Street (see also chapter 3, 'Development and Delusions').

But this story takes another turn, when in August 1971 Richard Nixon, the then president of the United States of America, unilaterally closed the gold window and 'by allowing the dollar to float, Nixon essentially abolished the 4000-year-old monetary system ... He killed money'.[107] This statement is only partly true, as the fractional reserve banking system which originated in sixteenth century Europe (see chapter 4, 'How History Accelerated') introduced a drastic modification to the old system – if it did not kill money it significantly corrupted it. However, economic theory has it that money has four functions: a unit of account, a medium of exchange, a store of value, and a means of settling accounts. But, 'without a link to the real world of gold and commodities, it was no longer a very good store of value'.[108] Chaos reigned in the global financial markets for a few weeks after this event and eventually settled by default into the floating exchange rate system which is in operation to this day. This system adversely effected the economies of the so-called third world countries, who were already at a disadvantage, by ensuring the steady devaluation of their currencies in relation to the developed countries. Hegemony rules.

About the same time as the old commodity-backed money lay dying, megabyte money was born with the advent of the computer, which 'drastically accelerate[d] its movements around the world ... [and] the number of dollars in existence could grow almost without limit';[109] minus one of the four elements that defined money in classical economics: it has no intrinsic value, it is simply computer code and thus wide open to abuse. George Soros, the Hungarian-American financier, showed us the vulnerability of this system. 'Soros [became] the Man Who Broke the Bank of England [when he] placed a $10 billion wager that Britain would be compelled to devalue the pound sterling. When this came about, on September 16 [1992], his funds were suddenly around $1 billion richer'.[110] This was a perfectly legitimate exercise although we might wonder about its scale. But it was also open to abuse by the unscrupulous, to the extent that in 2008 the world experienced an unprecedented banking crisis which led to a global recession from which economies are still limping towards a recovery:

> [O]r they are as shadows upon a sea obscure
> covered by a billow, above which is billow, above which are clouds,
> shadows piled one upon another;
> when he puts forth his hand, well-nigh he cannot see it ... (24: 40)

This then is the house of cards we inhabit, built layer upon layer with what modernity represents. The first layer consists of the banks which excretes phantom money that is the cement that holds this fragile edifice together. On top of which is built the nation state, industry, trade, business large and small, education, health, housing and social services, communication and all the multifarious institutions that keep our lives moving, all depending on the individual as a consumer and a source of tax at the top of this rickety edifice. Mervyn King, the ex-governor of the Bank of England, reflects: 'it is the young of today who will suffer from the next crisis – and without reform the economic and human costs of that crisis will be bigger than last time'.[111] King speaks as an economist. An environmentalist would add that one of the consequences of the lifestyles we have constructed for ourselves is to leave behind a degraded Earth as a legacy that our children will inherit. If we don't change now we will uniquely be leaving behind an enormity that can only be described as an intergenerational injustice on the grandest of scales.

Endnotes

1. JM Roberts, *The Pelican History of the World* (London: Pelican Books, 1980), p. 602.
2. Glenn J Ames, *The Globe Encompassed: The Age of European Discovery, 1500–1700* (Upper Saddle River, NJ: Pearson Prentice Hall, 2008), p. 4.
3. Updates from the Center for Earth Ethics, 'Respectful Request to Pope Francis to Rescind the Doctrine of Discovery, 1493' (YouTube: Somedayfire, 6 October 2017). Available at: https://www.youtube.com/watch?v=YM8yplamdLM (accessed May 2018). See also, '*Inter caetera* by Pope Alexander VI (May 4, 1493)' (Encyclopedia

Virginia, 20 March 2013). Available at: https://www.encyclopediavirginia.org/Inter_caetera_by_Pope_Alexander_VI_May_4_1493 (accessed May 2018).

4. Roberts, *The Pelican History of the World*, p. 601.

5. Ibid. p. 606.

6. Benny G Setiono, *Tionghoa dalam Pusaran Politik [Indonesia's Chinese Community under Political Turmoil]* (Jakarta: TransMedia Pustaka, 2008), pp. 111-113 (in Indonesian).

7. Ames, *The Globe Encompassed*, p. 176.

8. Gerald C Hickey, Joseph Buttinger, et al., 'Vietnam: The conquest of Vietnam by France' (Encyclopaedia Britannica, 26 July 1999–5 April 2018). Available at: https://www.britannica.com/place/Vietnam/The-conquest-of-Vietnam-by-France (accessed on 1 May 2018).

9. 'Commodore Perry and Japan (1853–1854)' (*Asia for Educators*, 2009). Available at: http://afe.easia.columbia.edu/special/japan_1750_perry.htm (accessed on 1 May 2018).

10. For the record the first 'circumnavigation' of the globe was undertaken by Ferdinand Magellan the Portuguese sailor, when after sailing west he entered the Pacific Ocean in 1620. He landed on the island of Cebu which was soon to become the Spanish colony of the Philippines. The essential difference between this and the American thrust Westwards was that an aggressive American capitalism was looking for markets for its industrial goods.

11. Carroll Quigley, *Tragedy and Hope: A History of the World in our Time* (New York: Macmillan, 1966), p. 7.

12. Op cit. p. 8.

13. The Earl of Cromer, *Modern Egypt*, two volumes (London: Macmillan, 1908), vol. II, p. 565.

14. Roberts, *The Pelican History of the World*, p. 600.

15. Op cit. p. 601.

16. Op cit.

17. 'AD 1493: The Pope asserts rights to colonize, convert, and enslave' (Native Voices, undated). Available at: https://www.nlm.nih.gov/nativevoices/timeline/171.html (accessed on 20 May 2018).

18. Roberts, *The Pelican History of the World*, p. 613.

19. Quoted in Peter L Bernstein, *The Power of Gold: The History of an Obsession* (New York: John Wiley and Sons, 2000), p. 126.

20. Roberts, *The Pelican History of the World*, p. 608.

21. South East Asia Study Resources, 'Minute by the Hon'ble T. B. Macaulay, dated the 2nd February 1835', Minute no. 10 (Columbia University, undated). Available at: http://www.columbia.edu/itc/mealac/pritchett/00generallinks/macaulay/txt_minute_education_1835.html (accessed on 1 May 2018).

22. Op cit., minute no. 33.

23. Op cit., minute no. 34.

24. John Newsinger, *The Blood Never Dried: A People's History of the British Empire* (London: Bookmarks Publication, 2006), p. 54.

25. Op cit. p. 86.

26. Op cit.

27. Quigley, *Tragedy and Hope*, p. 247.

28. Op cit. p. 247.

29. Op cit.

30. This is a reference to the assassination of Osama bin Laden, the founder and head of al-Qaida, in Pakistan on 2 May 2011, by a United States Naval Special Warfare Development Group.

31. Karl Mathiesen, 'What's the environmental impact of modern war?' (*The Guardian*, 6 November 2014). Available at: https://www.theguardian.com/environment/2014/nov/06/whats-the-environmental-impact-of-modern-war (accessed April 2018).

32. Roland Pease, 'Cern considers building huge physics machine' (BBC News, 18 February 2014). Available at: https://www.bbc.com/news/science-environment-26250716 (accessed on 11 June 2018).

33. 'Large Hadron Collider' (Science and Technology Facility Council, undated). Available at: https://stfc.ukri.org/research/particle-physics-and-particle-astrophysics/large-hadron-collider/ (accessed on 20 May 2018).

34. Samuel Miller McDonald, 'Extinction vs. Collapse: Does it matter?' (Millennium Alliance for Humanity and the Biosphere [MAHB], 8 May 2018). Available at: http://mahb.stanford.edu/blog/extinction-vs-collapse/ (accessed on 20 May 2018).

35. Described as the most important European philosopher of modern times. See 'Kant, Immanuel', in Ted Honderich (ed.), *The Oxford Companion to Philosophy* (Oxford: Oxford University Press, 1995), p. 434.

36. Op cit. p. 339.

37. Op cit. p. 619.

38. John Ralston Saul, *Voltaire's Bastards: The Dictatorship of Reason in the West* (New York: Vintage Books, 1993), p. 15.

39. Max Horkheimer in *Critique of Instrumental Reason* (Seabury Press: New York, 1974), quoted in Alain Touraine, *Critique of Modernity* (Oxford: Basil Blackwell 1995), p. 152.

40. Honderich, *The Oxford Companion to Philosophy*, p. 75.

41. Lynn White Jr, 'The Historical Roots of Our Ecologic Crisis', *Science*, New Series, Vol. 155, No. 3767 (Mar. 10, 1967), pp. 1203–1207. Available at: https://www.jstor.org/stable/1720120?seq=1#page_scan_tab_contents (accessed on 13 July 2018).

42. Antony Flew (ed.), *The Dictionary of Philosophy* (London: Pan Books, 1979), p. 204.

43. Richard Tarnas, *The Passion of the Western Mind: Understanding the Ideas That Have Shaped Our Worldview* (London: Pimlico, 1996), pp. 276–281.

44. Op cit. p. 319.

45. Robert Hamilton, *Earth Dream: The Marriage of Reason and Intuition* (Bideford: Green Books, 1990), p. 29.

46. Op cit. p. 30.

47. Touraine, *Critique of Modernity*, p. 1.

48. Theodore W Adorno and Max Horkheimer, *Dialectic of the Environment* (London: Verso, 1997), p. xiii.

49. Op cit. p. 42.

50. Claudio Corradetti, 'The Frankfurt School and Critical Theory' (Internet Encyclopaedia of Philosophy (IEP), undated). Available at: http://www.iep.utm.edu/frankfur/ (accessed on 1 May 2018).

51. 'What is Islamophobia' (Muslim Council of Britain, undated). Available at: http://www.mcb.org.uk/islamophobia/ (accessed on 2 May 2018).

52. They 'defended and upheld the transmitted beliefs of the Qur'an and Sunna [practise of Prophet Muhammad], as understood by mainstream Sunni Islam in each generation before them, from the extremes of excessive literalism and excessive rationalism.' See 'The Ash'aris & Maturidis: Standards of Mainstream Sunni Beliefs' (SeekersHub: Global Islamic Seminary, 19 November 2009). Available at: http://seekershub.org/ans-blog/2009/11/19/the-asharis-maturidis-standards-of-mainstream-sunni-beliefs/ (accessed on 2 May 2018).

53. Neal Robinson, 'Ash'ariyya and Mu'tazila' (Islamic Philosophy Online, undated). Available at: http://www.muslimphilosophy.com/ip/rep/H052 (accessed on 2 May 2018).

54. Arthur J Arberry, *Revelation and Reason in Islam* (London: George Allen and Unwin, 1957), p. 12.

55. *The Oxford Companion to Philosophy*, p. 446.

56. Op cit. p. 269.

57. Honderich, *The Dictionary of Philosophy*, p. 181.

58. Op cit. p. 70.

59. Op cit. p. 312.

60. Op cit. p. 182.

61. Arberry, *Revelation and Reason in Islam*, p. 18.

62. Yossef Rapoport and Shahab Ahmed (eds), *Ibn Taymiyya and his Times* (Karachi: Oxford University Press, 2010), p. 334.

63. Interpretation of problems not covered by the Qur'an or the actions of the Prophet Muhammad involving a process of reasoning usually undertaken by Islamic scholars.

64. Honderich, *The Dictionary of Philosophy*, p. 23.

65. Op cit. p. 24.

66. Roberts, *The Pelican History of the World*, p. 511.

67. See the chapter on the Natural World in Ismail Faruqi and Lois Faruqi, *The Cultural Atlas of Islam* (New York: Macmillan, 1986), p. 322.

68. Donald R Hill, *Islamic Science and Engineering* (Edinburgh: Edinburgh University Press, 1993), p. 66.

69. Ehsan Masood, *Science and Islam: A History* (London: Icon Books, 2017), p. 138.

70. Nils-Bertil Wallin, 'The History of Zero: How was zero discovered?' (YaleGlobal Online, 19 November 2002). Available at: http://yaleglobal.yale.edu/history-zero (accessed on 2 May 2018).

71. White Jr, 'The Historical Roots of Our Ecological Crisis'.

72. Op cit.

73. The Bible, *Genesis* 1: 128.

74. Quoted in Tarik M Quadir, *Traditional Islamic Environmentalism: The Vision of Seyyed Hossein Nasr* (Lanham, MD: University Press of America, 2013), p. 14.

75. Lynn White Jr, op cit.

76. John Gray, *Heresies: Against Progress and Other Illusions* (London: Granta Books, 2004), p. 2.

77. George Monbiot, 'The UK government wants to put a price on nature – but that will destroy it' (*The Guardian*, 15 May 2018). Available at: https://www.theguardian.com/commentisfree/2018/may/15/price-natural-world-destruction-natural-capital (accessed on 11 June 2018).

78. 'Chernobyl at 25th anniversary: Frequently Asked Questions' (World Health Organization, 23 April 2011). Available at: http://www.who.int/ionizing_radiation/chernobyl/20110423_FAQs_Chernobyl.pdf (accessed on 2 May 2018).

79. Ari Beser, 'After Alarmingly High Radiation Levels Detected, What Are the Facts in Fukushima?' (*National Geographic*, 22 February 2017). Available at: https://voices.nationalgeographic.org/2017/02/22/after-alarmingly-high-radiation-levels-detected-what-are-the-facts-in-fukushima/ (accessed on 2 May 2018).

80. Roberts, *The Pelican History of the World*, p. 601.

81. See the Alliance of Religions and Conservation (http://www.arcworld.org).

82. "Amish: North American religious group' (*Encyclopædia Britannica*, 28 May 2018). Available at https://www.britannica.com/topic/Amish (accessed on 14 July 2018).

83. David Orr, 'What Is Education For? Six myths about the foundations of modern education, and six new principles to replace them', *In Context: The Learning Revolution*, 27 (Winter 1992), p. 52. Available at: http://www.context.org/iclib/ic27/orr/ (accessed on 2 May 2018).

84. Op cit.

85. Op cit.

86. Roberts, *The Pelican History of the World*, p. 688.

87. Margi Prideaux, *Global Environmental Governance, Civil Society and Wildlife: Birdsong After the Storm* (Abingdon: Routledge, 2017), p. 31.

88. Roberts, *The Pelican History of the World*, p. 564.

89. This statement is attributed to Shaykh Abdalqadir as-Sufi (aka Ian Dallas), spiritual leader and intellectual, during one of his discourses in the 1990s.

90. 'The Big Picture: The People vs America' (Al Jazeera, 16 February 2017). Available at: https://www.aljazeera.com/programmes/the-big-picture/2017/02/big-picture-people-america-170216063801575.html (accessed on 2 May 2018).

91. Jeremy Seabrook, *The Myth of the Market: Promises and Illusions* (Bideford: Green Books, 1990), p. 10.

92. Op cit.

93. Op cit. p. 11.

94. Op cit.

95. Op cit.

96. Adam Smith, *Wealth of Nations* (Ware: Wordsworth Editions Ltd., 2012). See chapter on the 'Origin and Use of Money', pp. 27–33.

97. Josh Ryan-Collins, Tony Greenham, Richard Werner and Andrew Jackson, *Where Does Money Come From? – A guide to the UK monetary and banking system*, 2nd Edition (London: New Economics Foundation, 2012) p. 8.

98. Op cit. pp. 139, 140.

99. Op cit. p. 53.

100. Op cit. p. 48.

101. Op cit. p. 22.

102. Dorothy Rowe, *The Real Meaning of Money* (London: Harper Collins, 1997), p. 382.

103. King, *The End of Alchemy*, p. 96.

104. Bernstein, *The Power of Gold*, p. 25.

105. For a history of early money see Glyn Davies, *A History of Money: From Ancient Times to the Present Day* (Cardiff: University of Wales Press, 1994), pp. 1–60.

106. Richard Peet, *Unholy Trinity: The IMF, World Bank and WTO* (Kuala Lumpur: Strategic Information Research Development, 2003), p. 33.

107. Joel Kurtzman, 'Money's Demise', *World Business Academy Perspectives*, 8/2 (1994), p. 23.

108. Op cit. p. 23.

109. Op cit. p. 26.

110. Brendan Murphy, 'Finance: The Unifying Theme' (*The Atlantic*, July 1993). Available at: https://www.theatlantic.com/magazine/archive/1993/07/finance-the-unifying-theme/305148/ (accessed on 2 May 2018).

111. King, *The End of Alchemy*, p. 370.

2

A RAVISHED EARTH

A Disconnected People

Are we culpable?

I wish I was writing a travel brochure giving people a taste of the wonders of the natural world and its stunning variety; relishing the experience of describing the awesome beauty of creation, from the wonders of the oceans to the diversity of the rain forests to the grandeur of the mountains. I also wish that present day reality was such that it was possible for me to have chosen a different heading for this chapter, like 'A Ravishing World,' perhaps. 'Nature, in its ministry to man, is not only the material, but is also the process and the result'.[1] People more knowledgeable than I have written about the glories of the Earth with great sensitivity and will continue to do so with increasing regularity as they come to sense the continuing decline of the astonishing beauty and variety of the natural world. But what can one do when the sky is about to fall on our heads? How can one find solutions if problems get continuously brushed under the carpet? How do we face the enormous challenges that now confront us? The conundrum is that culpability for bringing the Earth to its knees rests, to a lesser or greater degree, with all of us. Uniquely, we are collectively the problem in the current frame of history. I cannot bring myself to believe that Homo sapiens will continue to ravish the Earth in its attempt to create a technotopia for itself and thus drive us further away from the bosom of the natural world that nurtures us:

The world is sweet and verdant
and verily Allah has made you stewards in it
and He sees how you acquit yourselves[2]

Good news concerning the environment is in short supply. As I was revising this chapter, three items of news appeared this week alone (mid-October 2017) that made me want to change the heading to something grimmer. Bad news is difficult to digest, and there is much of it around, but engaging the reader and getting to the point is important in this area, so here is the bad news from that week:

- 18 October: A study in Germany found 'a seasonal decline of 76%, and mid-summer decline of 82% in flying insect biomass over the 27 years of study. ... this decline is apparent regardless of habitat type, [and] changes in weather, land use, and habitat characteristics cannot explain this overall decline'.[3]
- 19 October: The *Lancet* Commission on Pollution and Health published a report reminding us that 'pollution is the largest environmental cause of disease and death in the world today, responsible for an estimated 9 million premature deaths'.[4]
- 23 October: An eight-year study has concluded that 'Plastic pollution, overfishing, global warming and increased acidification from burning fossil fuels means oceans are increasingly hostile to marine life'.[5]

So, how to be optimistic about this? These reports deal with the outcomes of human activity resulting from the lifestyles we have chosen and it is unlikely, for example, that those going on cruises will cancel their bookings on the basis of information on ocean acidification. It is more likely that these good people are blithely unaware of the connection between their desire for a luxury holiday and acidifying oceans. Then what of us who make a regular habit of flying, and others whose lifestyle choices mean driving endless miles, creating the pollution we are attempting so desperately to abate? One way out of this dilemma is to discover for ourselves that everything we do is connected in one way or another to everything else. This is a good place to start looking for a solution; no human is an island, and if you think you are, you still cannot escape the tides.

Accelerated human activity in recent times is disrupting the balance of the Earth's ecosystems; and the brunt of the consequences of our profligacy will be inherited by future generations:

Transgress not in the balance (mīzān),
and weigh with the justice, and skimp not in the balance.
And Earth: He (Allah) set it down for all beings (55: 8–10).

It only takes a few moments thought to realize that almost everything we do is energy intensive. From the moment I compose these words under an electric light, using a computer which has gone through a complex design, manufacturing, distribution and marketing process; from the production of the manuscript and the publishing, distribution and marketing that are essential to bring this book to you; reading it under an electric light, sitting comfortably in a chair which has been through these very same processes. And one could extend these thought processes ad infinitum. At every step in the book's journey we have used energy, burning non-renewable fossil fuels (atomic power and renewable energy still only supply a very small fraction of our energy needs) without so much as a thought to how all this was possible. And after a little more rumination, we realize that every single thing we need and want we take from the natural world. This kind of thinking could lead us to the inevitable conclusion that every single living human being imposes a demand on the natural world to a lesser or greater degree, and therefore is part of the problem if these demands are disproportionate and go much beyond basic needs. By the same logic we could also be part of the solution, by reducing our exaggerated needs and showing more consideration to the natural world.

Past generations lived simpler lives using far fewer resources than we do today, which meant that the demands they made on the natural world – our environment – were minimal resulting in an equally low energy demand. From about half a billion people in the mid-eighteenth century, when the human race was about to launch itself into the Industrial Revolution, there are now more than 7.6 billion of us and growing, individually making demands on the environment that were unheard of by people living in the past. Thus we have inflicted a double whammy on planet Earth: Alarming population growth coupled with exponentially increasing resource demands (see chapter 4, 'How History Accelerated' for a discussion on the exponential factor), and we continue this process unabated, disconnected from the very source that nurtures us. If this fact alone does not make us sit up and take notice then nothing ever will, unless of course we become victims of the consequences of our own actions. Uniquely, victim and aggressor are one and the same, and this tells us something about how we are running our world today. But what we are leaving behind for our children, and the enormous burden of this legacy, should trouble our collective conscience sufficiently to think and act differently.

In the summer of 2015 we were treated by the US-based NASA (National Aeronautics and Space Administration) to an image of planet Earth taken from outer space. We are familiar with images taken from Earth orbiting cameras, but this one was the first view of the entire sunny side of the planet from a deep space camera a million miles away. What we see is a delicate miracle suspended in space. It rotates on a finely defined axis set at an optimal distance from the Sun. It provides us with a barrier, the ozone layer, protecting us from dangerous cancer-inducing radiation. It gives us an atmosphere that allows us to breathe a finely-balanced mix of gases, providing us with the right proportion of carbon dioxide in the atmosphere to give us a habitable climate.

The processes that changed a hot molten ball, when the Earth first came into existence, to the one that looks like the NASA image has taken 4.6 billion years. The climate that enabled human societies to evolve into civilizations was made possible when the glaciers of the last ice age retreated 11,500 years ago. The processes that went into creating the kind of technology that enables us to look at our home in this manner have taken a fraction of a second by comparison. When we have recovered from the euphoria and the self-congratulation in its aftermath we may consider the warning that is embedded in this event. This product of great scientific and technological endeavour, laudable as it may seem, turns out to be a mirage; the inevitable outcome of a civilization trapped in a model of advancement that sabotages its own future.

A closed system

Climate change, or global warming, is nothing new in the history of the Earth's evolution, and in the past it would have been caused by variations in solar energy, volcanic eruptions, in the tilt or orbit of the Earth around the Sun – referred to as Earth Wobble – changing oceanic circulation and melting permafrost releasing large quantities of methane gas.[6] When the last glacial period ended, about 11,500 years ago, it opened the door to the kind of world we experience today and which created the conditions that allowed the human species to transform itself from being hunter gatherers to farmers. The deliberate planting of seed during one season in the year and harvesting the crop in another may seem commonplace to us today, but this aptly named Neolithic Revolution emerged about 10,000 years ago. Food self-sufficiency led to the growth of civilizations and urban centres, and they emerged about 8,000 years ago in what is now Mesopotamia, the area covered by present day Iraq, Syria and southern Turkey. Thus began the demand for energy, and it shouldn't be too hard to guess that at this point in our history it was entirely renewable. It remained that way all those thousands of years until about 250 years ago, when fossil fuels in the form of coal entered the energy equation on a massive scale and inaugurated the Industrial Revolution.

Another way of looking at this is to consider the Earth as a closed system. Everything life has ever needed through aeons of time has been made available to us by this system. Perversely, we can understand this better today using the technology we have created and considering how astronauts can live in space in small capsules over long periods of time. Scientists have copied the Earth's self-contained survival system and applied it to space travel. In this sense planet Earth could be likened to a spaceship. The most visible and commonplace example of this closed system is the rain cycle, and another system that is crucial to life on Earth is the carbon cycle. It maintains a climatic balance (*mīzān* in the parlance of the Qur'an) by regularizing the flow of carbon dioxide (CO_2) between the terrestrial, oceanic and atmospheric domains. There were natural changes to the carbon cycle in the past that induced climate change, but human-generated CO_2 since the Industrial Revolution has disrupted the carbon equation once again.

Carbon dioxide is a heat trapping gas, and excessive amounts of it pumped into the atmosphere by human action have created the greenhouse effect. To get to where we are now as an industrial consumer civilization we have been releasing this gas into the atmosphere at an ever increasing rate for the past 250 plus years, burning carbon in the form of coal, oil and gas that were locked in the bowels of the Earth by geological processes that took millions of years to form. This has resulted in the drastic increase of atmospheric CO_2, from estimated levels of 280 parts per million (ppm) during pre-industrial times and possibly for 20 million years before that, to 400 ppm today and growing. These anthropogenic (human induced) changes have confirmed our place in the geological calendar as a force of nature.

In addition to human brawn, it was not very long ago that we relied entirely on animal power and natural sources of energy such as wind and water to travel, move and lift objects and drive mechanical devices. We trod on the Earth lightly. As civilizations advanced we wanted more than the basics of food, clothing, and shelter and thus multiplied the demands we made on the natural world, making corresponding demands on Earth's energy resources. At the beginning of the Industrial Revolution we began to harness and perfect the power of steam to run our factories, trains and later, ocean liners. This required the energy supplied by coal, wood and any combustible carbon, like peat. By the end of the nineteenth century we had discovered the motivating force of electrical power and also oil as another source of energy in plentiful supply. The invention of the internal combustion engine paved the way for the transport revolution, with the mass production of the motor car and the expansion of road networks. Energy became available in our homes at the flick of a switch as electricity was harnessed to deliver power and light, thanks to the invention of the light bulb and other consumer conveniences like the vacuum cleaner, the washing machine, the fridge and much else. In the early twentieth century we invented heavier than air machines that could fly, which paved the way for mass air travel. All these inventions and discoveries that make life comfortable for us today and that we take for granted could only have been made possible by the extraction, transportation and consumption of vast quantities of fossil fuels. But, by the end of the twentieth century we had begun to discover the devastating effect their use was having on the environment.

> *Do not strut arrogantly on the earth.*
> *You will never split the earth apart*
> *nor will you ever rival the mountains in stature* (17: 37).

Geological epochs are usually measured in millions of years. The previous epoch, the Pleistocene, lasted about 2.6 million years and we are now just about living in the Holocene, which is estimated to have started about 11,500 years ago – a mere fraction of past epochs – which Homo sapiens has managed to disrupt. It would seem that we have given geologists something to think about: 'It's literally epoch-defining news. A group of experts tasked with considering the question of whether we have officially entered the Anthropocene – the geological age characterized by humans' influence on the planet – has delivered its answer: yes.'[7] We have become a force of nature in the twinkling of an eye. A very late-comer to planet Earth, an upstart species with an enlarged brain has contrived to shorten its welcome by subverting the very conditions of its own survival.

Although we no longer argue about the configuration of the Earth, we conduct our affairs today as if we still believe it were flat. We also appear not to be able to recognize the inter-connectedness of all of things, animate and inanimate, to each other and ourselves. Once we move past these barriers that shield our vision we come to the inevitable conclusion that our actions in one part of the world affect other people living on the other side of the planet, as the Earth is a closed system. As technology shrinks the Earth it appears to create a corresponding narrowing of our vision that alienates us from the natural world. The greenhouse gas (GHG) emissions[8] which cause global warming refuse to confine themselves to those industrial regions of the Earth where they were originally generated; there are no distant corners of the Earth any longer. The nature of nature is such that the stratospheric jet stream and planetary climate systems simply do not recognize national boundaries, and they tend to carry GHGs willy-nilly to all parts of the Earth. This then becomes a global problem, because the Solomon Islanders tucked away in the vast Pacific Ocean, who have contributed very little by way of GHG emissions, find themselves overtaken by a warming sea as a direct consequence of the buildup of these gases generated in industrialized countries.

Climate change

The feeling that something was amiss with climatic conditions as a result of human-induced factors began to emerge in scientific and technical circles in the early 1970s. The watershed for this movement was the conference held in Villach, Austria from 9–15 October 1985, attended by the World Meteorological Society, the United Nations Environment Programme (UNEP) and the International Council of Scientific Unions. The conclusion of the participants at the conference was that they could anticipate an unprecedented rise of global mean temperature in the first half of the twenty-first century. The scale and actual increase in global mean temperature was expected to be higher than any rise in the record of the planet's history. To mitigate the perceived events, the participants recommended a strategy that relied on technical and science-based research to establish target emission or concentration limits. In doing so, they sought to regulate the rate of change of global mean temperature within specific parameters.[9]

It is commonly held that the climate change issue moved from the realm of science to the realm of politics in the 1980s, and since then it has been high on the international policy-making agenda. The Intergovernmental Panel on Climate Change (IPCC)[10] was established in 1988 and climate negotiations were convened in 1992 to find an equitable solution to this problem. Since then, much of the delay in coming to an agreement has been caused by the reluctance of the advanced industrial countries to acknowledge fully the part they have played since the Industrial Revolution in creating this unfolding calamity. They have been behaving like selfish flat-earthers, and this mentality has pervaded climate change discussions since the international climate change treaty was negotiated at the Earth Summit in Rio de Janeiro in June 1992. The treaty, known as the United Nations Framework Convention on Climate Change (UNFCCC), came into force in March 1994. Its objective was to stabilize 'greenhouse gas concentrations in the atmosphere at a level that would prevent dangerous anthropogenic interference with the climate system'.[11] No binding limits were set on GHG emissions for individual countries and the framework contained no enforcement mechanisms, but it did outline how protocols could be negotiated that stipulated binding limits on GHG emissions.

The Conference of the Parties (COP), the governance mechanism of the UNFCCC, held its first session in Bonn, Germany in 1995. The original climate treaty, now familiar to us as the Kyoto Protocol, was agreed upon in 1997 and came into force in 2005.[12] The United States, the world's largest contributor of greenhouse gasses at the time withdrew from the protocol in 2001 and was the only nation to do so. It was due to be revised in 2012 but myopic national interests held this back. In the meantime, the Intergovernmental Panel on Climate Change (IPCC) reminded policy-makers in their 2014 Summary that 'Anthropogenic greenhouse gas emissions have increased since the pre-industrial era, driven largely by economic and population growth, and are now higher than ever.'[13] As Emily Adams of the sadly now defunct Earth Policy Institute observed:

> Despite wide agreement by governments on the need to limit emissions, the rate of increase ratcheted up from less than 1 percent each year in the 1990s to almost 3 percent annually in the first decade of this century. After a short dip in 2009 due to the global financial crisis, emissions from fossil fuels rebounded in 2010 and have since grown 2.6 percent each year, hitting an all-time high of 9.7 billion tons of carbon in 2012.[14]

The graph we see below, shown in Adams' article, bears a remarkable resemblance to one produced by another Adams, Henry Adams, at the birth of the twentieth century. The earlier graph deals with energy outputs in the nineteenth century, and I will discuss this further in chapter 4, but for now let us take a brief look at this one. It shows clearly the steep nature of the growth of global CO_2 emissions since the Industrial Revolution. What is more, the rate of

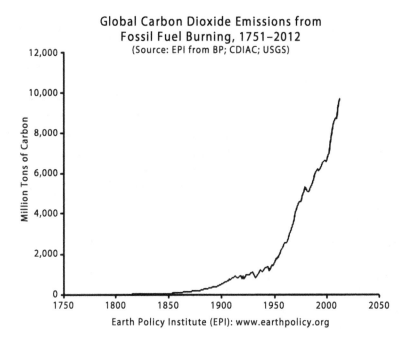

Global Carbon Dioxide Emissions from
Fossil Fuel Burning, 1751–2012
(Source: EPI from BP; CDIAC; USGS)

Earth Policy Institute (EPI): www.earthpolicy.org

increase grew at one and the same time as the international community began to show an interest in global warming, rising from around 6 billion tons in the early 1990s to 9.7 billion tons in 2012, the very same year in which the Kyoto Protocol was due to be replaced by another, more stringent, agreement. There were more signs here of the flat-Earth mentality, and a reluctance by politicians to act with the required alacrity to deal with the threats posed by global warming. Carbon dioxide output kept increasing, despite agreement on the need to limit emissions. The problem is systemic and is directly related to the economic growth agenda, an addiction from which economists and politicians find themselves incapable of freeing themselves.

However, there was a welcome shift in attitudes in 2015 when COP 21, also known as the Paris Agreement,[15] produced a compromise package that at the very least acknowledged that world leaders were beginning to see the writing on the wall. There were three major outcomes in Paris and they were to hold global warming to well below 2°C, with 1.5°C as a target figure to aim at, a collective goal for greenhouse gas emissions to peak as soon as possible and to achieve net-zero emissions in the second half of this century. There is also provision in the Paris Agreement for developed countries to take the lead in mitigating climate change and support the actions undertaken by developing countries. Alas the agreed support was wholly inadequate. And additionally, researchers "left COP26 dissatisfied at the lack of stronger commitments to reduce emissions, and failure to agree ' "loss and damage" finance for countries that are vulnerable to climate change' (Nature, 14 November 2021). All too late for the Solomon Islanders as they have already passed the point of no return as 'the edifice of international climate policy rests on a target that no one believes it is likely can be met.'[16]

So what is the legacy we are leaving behind for future generations? It is observed in the Fifth Assessment Report of the Intergovernmental Panel on Climate Change that, 'Continued emission of greenhouse gases will cause further warming and long-lasting changes in all components of the climate system, increasing the likelihood of severe, pervasive and irreversible impacts for people and ecosystems'.[17] Although the future is threatened with drastic consequences, the concern we need to demonstrate on behalf of the generations that will succeed us nowhere matches the enormity of what we face. Some analysts suggest that understanding the consequences of climate change is not within the reach of the experience of people unless something like the Solomon Island phenomenon directly affects their lives. Or is it all too big to comprehend? What is incomprehensible is the decision of the current American president to take his country out of the Paris Agreement, while at the same time the news gets worse. 'Greenhouse gas concentrations surge to new record' was the headline of a bulletin released by the World Meteorological Organization on 30 October 2017:

> Concentrations of carbon dioxide in the atmosphere surged at a record-breaking speed in 2016 to the highest level in 800 000 years ... The abrupt changes in the atmosphere witnessed in the past 70 years are without precedent,
>
> Globally averaged concentrations of CO_2 reached 403.3 parts per million in 2016, up from 400.00 ppm in 2015 because of a combination of human activities and a strong El Niño event. Concentrations of CO_2 are now 145% of pre-industrial (before 1750) levels ...
>
> Rapidly increasing atmospheric levels of CO_2 and other greenhouse gases have the potential to initiate unprecedented changes in climate systems, leading to 'severe ecological and economic disruptions'.[18]

There appears to be no end to human folly.

The Pollution Boomerang

The hole in the sky

We have become so disconnected from the natural world that we have become incapable of comprehending what we are doing to it and consequently to ourselves. Our actions in pursuing selfish short-term exaggerated states of comfort cause discomfort to all in the long term. There are too many of us now wanting much, much more than past generations. Contentment is now a scarce commodity. The price of this inflated sense of wellbeing is at the expense of humanity itself, not to speak of other sentient beings. We desecrate the earth we tread on as we reach out to grab the mirage of the good life which forever keeps eluding our grasp.

> *... their works are as a mirage in a spacious plain which a man athirst thinks it is water but when he reaches it, he finds it to be nothing; there indeed he finds Allah.* (24:39)

Take the motor car for instance, that most coveted of objects in our consumer civilization that ostensibly makes life so easy for us. Just over 40 million cars were produced in 2000 and over 72 million in 2016; that is at the rate of 197,500 cars each day.[19] By 2020, annual car production is expected to rise well above 100 million and 'It is estimated that over 1 billion passenger cars travel the streets and roads of the world today'.[20] The volume of CO_2 and other polluting gases that are being spewed into the atmosphere can hardly be imagined. The invention of the catalytic convertor was expected to solve this problem, but its effect is only partial. Contrary to popular opinion it does not deal with CO_2; it was not intended to, as CO_2 is not a toxic gas, but it is a pollutant nevertheless. As we have seen this is the underlying cause of global warming and it also contributes to the acidification of oceans. In addition to CO_2, car exhausts produce nitrogen, which is benign, water vapour and a cocktail of toxic gases.[21] These toxic gases are composed of smog-producing hydrocarbons that can also be the cause of asthma, liver and lung disease, and cancer; volatile organic compounds, a source of chlorofluorocarbons (CFCs) that deplete the ozone layer; nitrogen oxides that contribute to the occurrence of acid rain and can harm respiratory systems of animals and plants; remnants of sulphur dioxide, another contributor to acid rain, that legislative regulation has failed to remove; poisonous carbon monoxide gas; and benzene, which is toxic, carcinogenic and linked to leukaemia.

The catalytic convertor has so far met the requirements of existing clean air legislation in developed countries, but it is anticipated that it will not satisfy future legal requirements because of an expected steep rise in the volume of motor traffic. But there is another matter, and it is the catalytic converter itself. In order to achieve its objectives, these converters need certain rare metals like platinum, rhodium or palladium, and now increasingly, gold. These metals have to be mined, processed and then assembled into the final product. In short the converter is the end result of a mining and manufacturing process that itself causes further environmental aggravation. Much of the palladium and platinum is mined near the Russian city of Norilsk inside the Arctic Circle, the result of which is that it is now one the world's ten most polluted cities.

It may be argued that the advent of the electric car will change all this and as a result there could be both a meaningful reduction in global CO_2 output and environmental impact, as the catalytic convertor by then will be made redundant. This is true only up to a point, as someone somewhere has to produce the electricity that is needed to run the billion-plus cars already on the road. It is unlikely that the production of the required amount of electricity is going to be entirely through the use of renewable energy, because of the longish lead times needed for the changeover. By that time it will be too late anyway and we will then need to consider the impact the production of this number of cars is going to have on ecosystems. And what of the much-lauded lithium battery which is going to replace the petrol tank? Lithium is flammable and requires a chemical process to be made safe; its extraction and mining process uses large amounts of water; toxic chemicals are released during the manufacturing process; nickel and cobalt used in the production of lithium batteries cause further environmental hazards.[22] We are for sure going to leave fresh problems for the generation at our heels.

The invention of the refrigerator, freezer and air conditioners has propelled modern consumer lifestyles to another level. Spring lamb from New Zealand in the southern hemisphere can be eaten in the winter in northern countries; strawberries, a northerly summer crop, can now be grown in greenhouses any time, transported anywhere, stored for any length of time and eaten at the whim of the consumer. Eating imported strawberries in Lagos, Nigeria, is no longer an uncommon experience. Factory trawlers can fish in tropical waters, process their catch on board, and deliver it to their markets in distant northern ports weeks after they are caught. Rich oil countries in the Middle East can without a thought build glasshouse-like skyscrapers in their hot climate and use air conditioners to cool the exaggerated heat these structures create.

Refrigeration along with air conditioning is made possible by artificial inert gases that belong to the chlorofluorocarbon (CFC) family. But there is a sting in the CFC tail. It is ironic that the chemical stability that makes CFCs so suitable for use in household items such as refrigerators should be the cause of creating much global alarm. When released into the atmosphere, most chemicals get rapidly broken down into smaller, harmless components by reactions in the lower atmosphere. However, CFCs are so stable and non-reactive that they survive to reach the highest levels of the atmosphere, and become globally distributed in the stratosphere. At these high altitudes the intensity of ultraviolet (UV) radiation is so great that even the stable CFCs are split apart to release a chlorine atom.[23] It is this chlorine that sets about destroying the ozone layer, the thin region of gas found on average at an altitude of about 15 miles (25 km) above the surface of the Earth, in the lower part of the stratosphere. We don't usually give much thought to the fact that the Earth is a delicately balanced system, but by absorbing much (97–99 per cent) of the Sun's ultraviolet radiation this layer plays a vital role in protecting terrestrial life forms right down to the bottom rung of the food chain. Ultraviolet radiation can lead to skin cancers and vegetation damage. As the Qur'an reminds us: 'We (Allah) made the sky a well secured canopy – yet from its wonders they turn away' (21: 32).

The hole in the ozone layer was discovered above the Antarctic in the 1980s. This triggered an international reaction which led to the creation of the treaty known as the Montreal Protocol on Substances that Deplete the Ozone Layer.[24] It came into force in January 1989 and was designed to progressively reduce and ultimately eliminate the use of the CFC family of inert gases in manufacturing processes and replace them with less harmful agents. Scientists estimate that it will take another 50–100 years before the ozone layer is fully restored, but in seeking to find an alternative to CFCs we have created another problem. Much to everyone's delight at the time, scientists, technocrats and manufacturers came up with an answer that would ostensibly ensure the successful plugging of the hole in the sky. Thus Hydrofluorocarbons (HFCs) were born. But much to the chagrin of those who make these decisions for us, HFCs turned out to be a greenhouse generating gas on a collision course with the Kyoto Protocol responsible for monitoring GHG emissions that cause climate change. As Greenpeace International would have it, 'HFCs are often misleadingly portrayed by their manufacturers and users to be "environmentally friendly" because they do not deplete the ozone layer. However, they are highly potent greenhouse gases and contribute significantly to climate change'.[25]

The reaction of the international community to this setback was to produce the Kigali Amendment to the Montreal Protocol. The HFC Phase-down was ratified in Kigali, Rwanda, in October 2016.[26] But given the nature of international agreements, developing countries like China and India are not expected to phase out the use of these gases until the late 2020s and early 2030s, respectively. Thus the rising middle classes of these countries will be assured of their quota of ice cream even as the planet warms up.

People, poison and plastic

Globalization appears in various forms: multinational corporations scouring the earth for profit; the World Bank financing dams, destroying pristine environments in the process and disrupting the lives of people in the name of progress; container ships crisscrossing each other on the high seas as one carries cars from the east to west while the other does the same in reverse; the trend towards a global monoculture in fashion, food and music; the ubiquitous Coke and hamburger turning up in all corners of the Earth. Climate change is global: the result of millions of us acting individually but not really comprehending that we have all contributed to this mess to a greater or lesser degree, no matter where we live. Then there is acid rain, the result of burning coal to generate electricity. This activity produces waste gases which contain sulphur and nitrogen oxides that combine with atmospheric water to form acids that destroy forests and pollute lakes and waterways. The UK exported this kind of pollution until the last decades of the last century and was dubbed 'the dirty man of Europe' by the Scandinavians whose forests and lakes suffered as a result. The Chinese appear not to have noticed this and have been building coal-fired power stations at the rate of one a week. It is reported more than 250 of its cities are suffering from the effects of acid rain:

> 'Leshan Giant Buddha, which has stood in southwest China for more than 1,000 years, has also been badly affected. Its nose is turning black, hair curls have fallen from its head and its reddish body is becoming a charred grey colour' ... The 71-metre high and 28-metre wide statue, which was carved out of a cliff during Tang Dynasty (AD 618–907), had survived floods and earthquakes, but it was now at a greater risk from the man-made threat.[27]

The people of the Indian sub-continent are not far behind as the Taj Mahal too is being corroded as a result of the effects of toxic atmospheric discharges from factories in the region. There must be another way of delivering the good life to the people.

We are now encountering something entirely new. It unfolds in its own time, it is global and in all our homes, without exception, it is surreptitious and human-induced. It has been here ever since we began to concoct artificial chemicals and dispose of them without thought as to where they were going to end up. It's in the air, in the water and in the food chain. It ingratiates itself into our bodies, lingering there for as long as it wants and does its work by stealth. There is now a class of artificial chemicals known as endocrine disruptors, which are described as persistent organic pollutants (POPs). They are known to 'interfere with the body's endocrine system and produce adverse developmental, reproductive, neurological, and immune effects in both humans and wildlife'.[28] POPs are found in many everyday products – including plastic bottles, metal food cans, detergents, flame retardants, food, toys, cosmetics and pesticides. Research shows that endocrine disruptors may pose the greatest risk during prenatal and early postnatal development when organ and neural systems are forming.

I was at the State of the World Forum in San Francisco in 1996, the kind of gathering that heads of state such as Mikhail Gorbachev, the former leader of the Soviet Union, would attend. It was the turn of the late Theo Colborn[29] to take the stage. Cutting a slight figure, she looked the audience in the eye and stabbing her finger in the air said: 'Do you know each one of you have more than two hundred and fifty poisons in your body your grandparents did not have?' Al Gore, in his foreword to her book *Our Stolen Future*, says:

> *Our Stolen Future* provides a vivid and the readable account of emerging scientific research about how a wide range of manmade chemicals disrupt delicate hormone systems. These systems play a critical role in processes ranging from human sexual development to behavior, intelligence, and a functioning of the immune system.
>
> Although scientists are just beginning to explore the implications of this research, initial animal and human studies link these chemicals

to myriad effects, including low sperm counts; infertility; genital deformities; hormonally triggered human cancers, such as those of the breast and prostate gland; numerological disorders in children, such as hyperactivity and deficits in attention; and developmental and reproductive problems in wildlife.[30]

One gathers from this that no one is free from being contaminated by an alien chemical regardless of where one lives, even far from the source where the pollution emerges. People living in the remotest places on Earth are still contaminated, often to very high levels. The Inuit people living within the Arctic Circle, which is quite some distance away from industrial activity, have been known to carry high levels of persistent organic pollutants. These pollutants also travel deep inside the human body and contaminate the foetus as it forms; and also the newborn child through the milk of its mother. At a time when we are worrying about an overcrowded planet it may be that the solution is coming from ourselves. There is an increasing amount of evidence to suggest that endocrine disrupters are responsible for the rising levels of male infertility.[31] An insidious way of reducing population growth and taken along with other factors this trend is cause for much alarm.

In the long term the pollution that spills out from the lifestyles we have chosen eventually attacks our brain function. Much of everything we 'inhale, ingest, and absorb over a lifetime: air pollutants, smoke, gas fumes, pesticides, cleaning agents, aluminium, lead, PCBs,[32] iron, mercury. Many of these, such as pesticides and metals, are neurotoxins. Others trigger inflammation and oxidative damage that destroy brain tissue. And yes, experts do say that chronic exposure to environmental toxins can increase the risks of age-related memory impairment and dementia.'[33] In a news release launching two reports, the World Health Organization (WHO) wasted no time in emphasizing that children are affected by these pollutants on a massive scale:

> More than 1 in 4 deaths of children under 5 years of age are attributable to unhealthy environments. Every year, environmental risks – such as indoor and outdoor air pollution, second-hand smoke, unsafe water, lack of sanitation, and inadequate hygiene – take the lives of 1.7 million children under 5 years ... 'A polluted environment is a deadly one – particularly for young children,' says Dr Margaret Chan, WHO Director-General. 'Their developing organs and immune systems, and smaller bodies and airways, make them especially vulnerable to dirty air and water.'[34]

Who would have thought that a simple device like the household tap would connect us to the extreme depths of the oceans? Every day we unthinkingly wash and flush away the chemicals in our soaps, toothpastes, cosmetics, contraceptive pills, detergents and other household waste down the drain. As they wend their way towards the coast these chemicals join toxic industrial waste – mostly the result of manufacturing processes that turn out our consumer comforts – and it all ends up at the bottom of the ocean. And it is not just the ocean floor, it is the deepest part of the ocean one can find, described as the last major marine ecological frontier:

> The legacy and reach of anthropogenic influence is most clearly evidenced by its impact on the most remote and inaccessible habitats on Earth. Here we identify extraordinary levels of persistent organic pollutants in the endemic amphipod fauna from two of the deepest ocean trenches (10,000 metres). Contaminant levels were considerably higher than documented for nearby regions of heavy industrialization, indicating bioaccumulation of anthropogenic contamination and inferring that these pollutants are pervasive across the world's oceans and to full ocean depth.[35]

Are we turning the oceans into a consumer cesspit?

We can get closer to the answer by looking at the role the ubiquitous plastic plays in our lives. Without this creation of modern science our civilization would probably collapse. It occupies every niche in our lives, including our bodies, and it is a synthetic material valued for its longevity and light weight, which are exactly the qualities that are creating problems for us all. 'Of the 260 million tons of plastic the world produces each year, about 10 percent ends up in the Ocean,'[36] causing mayhem to marine life, sea birds, forming huge plastic islands in the sea and polluting beaches. On land it chokes drains, waterways and rivers. And in our bodies it is the source of endocrine disruptors that interfere with child development. What do we do about it?

> This unprecedented plastic waste tide appears as vast as the ocean, as ungraspable as the unfathomable mass of microscopic plastic fragments present at sea, transported by winds and currents, yet, ultimately, the plastic tide can become as limited as our chosen relationship with plastics, which involves a dramatic behavioral change on our part. The path to successful resolution of the crisis clearly appears ... as we are the problem and the solution.[37]

There has been piecemeal legislation in some advanced countries to prevent the manufacture of carcinogenic compounds and endocrine disruptors in certain varieties of plastics, and some countries have gone to the extent of banning the use of plastic bags altogether. But until someone discovers a means to degrade plastic or recycle it without causing further problems or synthesizing new materials to replace it (God forbid), we are stuck with plastics in timescales that can be counted in the hundreds, if not thousands, of years. We have created a substance whose nature is alien to nature. Until such time as we discover a once and for all solution, if there is ever going to be one, there needs to be a global protocol matching the one dealing with climate change to deal with the manufacture, use and disposal of plastics and a massive education programme directed at the general public to create awareness of this curse that human ingenuity has created.

Rape of the Forests

Plantations and cash crops

On one of my visits to Sri Lanka, the country of my birth, one of my brothers took me to the mountains where the tea plantations were established by the British. In a sea of tea bushes stood a lonely sentinel – one solitary tree lamenting the loss of its community. Then it struck me! The cost of a cup of tea was the decimation of mountain range after mountain range of pristine rainforest. The railway then arrived, driven by imported coal, travelling on imported rails placed on sleepers cut from logs that came from the rainforest. The railway track was built to go up to almost 6,300 feet (1,898 metres) into the mountains to ship the famous Ceylon tea, still a brand to be reckoned with. It paid the wages of the miners and steel workers in Britain more than 150 years ago.

This is typical of what happened during the period of European colonial expansion from Chile to China. And to tea we can add coffee, rubber, tobacco, cotton, sugar cane and those many more agricultural products that we call cash crops. Of course, deforestation did not end with decolonization. It is estimated that at the turn of the nineteenth century, 90 per cent of Sri Lanka was forested, and in spite of the vagaries of colonialism it still had about 50 per cent cover in the 1920s. By 2005, this had fallen to less than 25 per cent. During the twentieth century, forest cover in the Philippines went from 70 per cent to 20 per cent. Malaysians and Indonesians are destroying their rainforests to plant oil palm and biofuel crops, the Chinese are destroying theirs in Yunnan Province for rubber, and the Brazilians are destroying the Amazonian Rainforest for logging, mining, cattle ranching and planting tobacco and soya. A similar process is taking place in North America. In the 'United States deforestation has caused the destruction of virgin forests by 90 percent since 1600. At the rate of deforestation currently in the United States, only one-fourth of the forests standing today will be standing in 70 years'.[38] The British Isles was known to be fully wooded 5,000 years ago. It shrunk as the population grew and demand for land increased. The last wolf was seen in England in the fourteenth century.

It is estimated that 10,000 years ago about 35 per cent of the planet was covered by forest. Surprisingly, about 30 per cent of the planet remained under forest cover right up to the middle of the last century, and in this time all the great civilizations of the past and the Industrial Revolutions in Europe and North America had only consumed roughly 5 per cent of forest cover; some estimates even put this as low as 2 per cent. Yet it is now estimated that only 10 per cent of the planet's surface is forested today. We may conclude from these figures that in the last fifty years our modern civilization has destroyed more than four times the amount of forest resources than all other civilizations put together in the past 9,950 years.

As a species, human beings are inescapably interwoven into the fabric of the natural world. We are just one entity in creation amongst a myriad others and it doesn't belong to us, although we take from it copiously: *The creation of the heavens and the earth is far greater than the creation of humankind. But most of humankind do not know it* (40: 57). But as intelligent beings we have responsibilities and *'it is He (Allah) who appointed you as stewards on the Earth'* (6: 165).[39] As stewards we protect what gives us life and nurtures us by its abundance. The forests contribute to the oxygen cycle through photosynthesis. They act as carbon sinks and absorb vast quantities of it as part of the carbon cycle. They perpetuate the water cycle, contribute to the Earth's weather patterns and keep the forest floor stable by locking it in place with its roots. We were once integral to the forests and all of our needs came from them. We shared the plants and trees and their diversity with other beings. Is it merely a resource? Today this question is loaded and only a mindset that is economic in its outlook can ask it, as indeed environmental economists do, and the idea is catching on. A sacred entity is being monetized and in its journey from forest to consumer it will appear as computer code in bank records. It is of the same genre as asking if our hearts and lungs are resources. An economist might say yes to that too, and might even conjure up an economic theory to prove the point and perhaps put a monetary value on them. This idea has been around for a while and has been given added credibility by the Millennium Ecosystem Assessment. Over a thousand experts worldwide submitted 'a state-of-the-art scientific appraisal of the condition and trends in the world's ecosystems and the services they provide'.[40] If the forests could speak they would probably tell us that they are humiliated by the idea because their value to us is beyond measure: *'And the plants (stars) and trees prostrate themselves (to the Creator)'* (55: 6).[41]

Subsistence farming has been a way of life for millennia. The scale of this activity in the past was such that it ensured forest recovery and revival. Now population growth and poverty has combined to ensure that this activity is classified as one of the major causes of deforestation. This is now causing concern in Brazil, as the Amazonian Rainforest is systematically encroached. Indonesia finds itself in the same predicament and it is now a political football. Using wood from the forests for fuel is another cause for concern and again it is a matter of scale. What else would the poor use for cooking? But massive areas of forest are being stripped clean or put to the torch for commercial purposes, the most prominent being palm oil and sugar cane, the by-products of which appear on our supermarket shelves. They are also the sources for biofuels that are ostensibly meant to replace fossil fuels.

Carbon sink

If a herd of elephants strip a forest bare they move to another location allowing the denuded area to regenerate. If they have nowhere else to go that spells the end of the herd. Ultimately in nature there is a balance between a species' needs and its numbers, and if this is upset then there is a natural cull. This happens to human beings too and like the Easter Island episode (see chapter 3, 'The Progress Trap') it is not all that uncommon.[42] We apply our intelligence to find ways out of tight corners and our ingenuity and adaptability has ensured our survival. The conventional answer has been to move on to fresh pastures and adapt, but we now populate every niche on the planet and our rapid growth in numbers is testing our ingenuity to the extreme. Technology is offered as a fix, but new technology habitually brings us new problems as we have seen in the case of the catalytic convertor, and in new materials, as in the case of plastics. In the meantime, let us not forget what caused global warming in the first place.

Paradoxically, the forests were safest when our ancestors lived in them as hunter-gatherers. They understood the forest, took from it what they needed and left the rest alone. The relationship began to change when they turned to agriculture about 8,000 years ago. They learnt to make charcoal and used wood to make shelters and homes as they emerged into the open. Slash and burn was a primitive form of encroachment agriculture practised at the forest's edge. Agriculture took us one step away from the forests, industrial society another, and technological modernity yet another. It would seem that the more we distance ourselves from the forests, the more we seem to want from them and the more we do to destroy them in the process. A disconnection has taken place, and links in the chain that anchored us to the natural world have been broken. We have created artificial spaces for ourselves at the expense of what is natural as our alienation from nature intensifies.

As we have seen, since the Industrial Revolution began in the eighteenth century there has been a drastic rise in carbon dioxide concentration in the atmosphere and, 'Loss of forests contributes between 12 percent and 17 percent of annual global greenhouse gas emissions.'[43] Forests are carbon sinks – they absorb CO_2 into themselves. A fully grown tree locks anything between 10 to 15 kilograms of carbon dioxide within itself. In the narrow context of greenhouse gases, we are causing a double detriment to ourselves. We pump carbon dioxide into the atmosphere at alarmingly increasing rates and destroy the very resource that can absorb it. This is a very complicated and painfully slow way of strangling ourselves to death – as we rapaciously destroy the forests we are also causing detriment to the very source of the systems that keep us alive. The sensible thing to do would be to reverse one or preferably both of these destructive processes, and it is highly unlikely that the COP proposals are going to save us from the destructive impact of climate change. Reforestation schemes lag way behind the rate of destruction and it is also useful to know that it takes anything between twenty to fifty years for a tree to grow to full maturity to fulfil its potential. We need to put an immediate stop to the destruction of forests and reclaim them at an accelerated rate to get back to some kind of equilibrium. Proposed carbon sequestration programmes are laborious and untested techno-fixes, and they can in the long term best be seen as an adjunct to reforestation programmes and a progressive conversion to the use of renewable energy.

Sustainable development in this context is a palliative, and if we do not act with the required urgency a political and economic model that dazzles us with its consumer culture will drag us down to the bottom as it did the *Titanic* when it hit the iceberg.

Ecocide

Bio diversity

When I caught a cold as a child, my mother made me a concoction based on coriander. I remember it did me a lot of good, and this infusion would appear in various forms mixed with other herbs, depending on what I was complaining about. Plant-based remedies are very much a part of the human story and still a part of the lives of many, even in developed societies, although modern medicine has tended to elbow them out. It may be that the popularity of the herbal teas we consume is our way of keeping in touch with what we are losing. Although the production of the ubiquitous Aspirin is now entirely synthetic, its origins are plant-based. Salicylic acid, the active ingredient in aspirin, is found in willow bark, meadowsweet and Spiraea and has been known to cure headaches, pains and fevers. The Madagascar rosy periwinkle has been used by traditional healers to treat diabetes and it is now the source of medicines used to fight leukaemia, a cancer of white blood cells. Although modern medicine has taken over it would seem that plant-based remedies ensured the flourishing and survival of humanity until recent times.

In ages gone by our ancestors experimented with all the parts of plants: their leaves, flowers, fruit, nuts, sap, bark, roots and whatever else they could squeeze, shred or extract out of them. They will have tasted sweet, sour, hot and bitter. The results would have been to satiate their hunger, quench their thirst, drug them, cure them and sometimes kill them. This was their long-term longitudinal experiment in the vast laboratory of life. We could not have survived without the biological diversity of the forests. Diversity is the nature of nature. Although we extract the most benefit out of the Earth's gifts, they are not exclusively ours and other sentient beings in nature have an equal right to the benefits nature bestows on us. The difference between us and them is that we take what we want while they only take what they need.

Deforestation is the primary cause of the loss in biodiversity. As we have seen we are losing huge tracts of forest cover every day and it has been estimated that we destroy an area the size of Panama each year. The World Wildlife Fund estimates that some 46–58,000 square miles of forest are lost annually; an area equal to the size thirty-six football fields every minute.[44] We are ravaging the legacy that is our responsibility to leave behind for future generations. There is a strange schizophrenia about this matter. As we destroy the natural world we also commercialize its treasures, as pharmaceutical companies and university research departments scour the forests for their secrets. We knowingly extinguish what is of value to us while we benefit from its vanishing remnants. It is symptomatic of our modern society that we profit from destruction.

The picture our minds conjure up when we think of deforestation is of a crashing tree after it has been humiliated by a chainsaw. The felled tree is severed from an ecosystem[45] of which it was an integral part. The trees and plants that grow together have developed synergies between them, along with the fauna that depend on and live off them. A tree is valuable, from its highest branch to its lowest roots; from harnessing the sun's energy in chlorophyll to produce glucose and the by-product, the essential (for us) oxygen, to locking CO_2 in its trunk and keeping the forest floor stable with its roots. There are life forms from its highest reaches to its deepest roots. When a forest is cut a thriving cosmopolitan nation dies, and as Tony Juniper, one-time head of Friends of the Earth, observes:

> While biodiversity is often measured and expressed in numbers of species, it is habitats that are the critical factor in determining the health of life on Earth. From the depths of all the oceans to the exposed peaks of high mountains, it is the wellbeing of natural environments that is most critical to the conservation of species and populations of animals and plants.[46]

In addition to mammals, reptiles, birds and bees, there are insects and microbes that assist in the breakdown of decaying matter and help to maintain the atmosphere, fungi that break down decomposing waste, and plant-like algae that produce much of the oxygen we breathe. We have given scientific names to about 1.5 million of them but we have no idea how many there are altogether and it could be anything from another 1.5 million to a 100 million. Our knowledge gap is also exposed as we have no idea how they contribute to the wellbeing of their habitat, to us and ultimately to the planet itself.

What the eye doesn't see the heart doesn't grieve over. We have come a long way, the argument goes: the human race has thrived for millennia without knowing the deeper secrets of the natural world, so why do we need to know now? It's there for us to take anyway; is enriching ourselves not more important than the forests and the oceans and the life forms that thrive in them? But, there is another side to the human psyche. Why do we keep pets and set up all sorts of organizations for their protection, welfare and retirement? Why do we go to the zoo? Why do we love the cuddly panda, the lissome tiger, the brooding gorilla? The last three named iconic animals, along with a host of others, are now on the lengthening list of endangered species, all the result of the human assault on their habitats.

Extinction

There is now a general consensus among scientists and wider afield that we are in the middle of the sixth mass extinction. Richard Leakey, palaeontologist and environmentalist, asserts: 'Dominant as no other species has been in the history of life on Earth, Homo sapiens is in the throes of causing a major biological crisis, a mass extinction, the sixth such event to have occurred in the past half billion years. And we, Homo sapiens, may also be among the living and dead.'[47] This is without precedent and species loss has accelerated since human intervention, conjuring up the possibility that the only spaces where our children will be able to see 'wildlife' are in the zoos and movies. The Qur'an observes:

> *Allah created every animal from water*
> *Some of them creep on their bellies,*
> *some that walk on two legs and some on four* (24: 45);

and,

> *There is no creature crawling on the earth,*
> *or those that fly,*
> *who are not communities like yourselves* (6: 38).

The following are estimates made by biologists of the five major extinctions in the pre-human era:[48]

Major extinctions	Approximately when	Approximate % of Species loss	Possible cause/s
First	450 million years ago	86%	Climate change.
Second	375 million years ago	75%	Changes in planetary ecology.
Third	251 million years ago	96%	Global warming, ocean acidification.
Fourth	200 million years ago	80%	Clear causes not apparent.
Fifth	66 million years ago	76%	Asteroid impact.

We know why the dinosaurs disappeared from the surface of the Earth in the fifth major extinction, but we don't have a full picture of what we are eradicating in the anthropogenic sixth mass extinction, because we don't know enough about what is out there. Comparisons with the past are always useful, even if the picture of the past is far from complete, as they can point us only in certain directions and enable us to come to reasonable conclusions about the trends we experience. Unlike in even the very recent past, we have discovered today a great deal more than our forebears discovered in their time. The further we go back in time the less certain we can be of events of this kind but the destructive tendencies of the human species become more evident as we get closer to our own times and are projected to get much worse in the future. Here is a list of human-induced extinctions starting from the fifteenth century, when we can be more certain of events. The list is biased in favour of mammals and birds but this is simply a reflection of us having more information about these species compared with others and we should be under no illusion that we are losing plants and insects, known and unknown, at a much greater rate.[49]

Notable human induced species extinctions		
Date	**Species (Region)**	**Causes**
*c.*1500	Mao (New Zealand)	Giant flightless bird. Hunted to extinction.
*c.*1600	Haast's Eagle (New Zealand)	Main predator of the Mao; a good example of species interdependence.
1662	Dodo (Mauritius)	Gave rise to the expression, 'Dead as a dodo'. Hunted to extinction.
1768	Stellar's Sea Cow (North Pacific)	Massive marine herbivore hunted to extinction 27 years after discovery by Europeans.
1800	Bluebuck Antelope	First African antelope to be hunted to extinction by European settlers.
1860	String Tree (St Helena)	Habitat destruction.
1886	Bennett's Seaweed (Australia)	Also known as red alga. Massive human activity.
1889	Hokkaido Wolf (Japan)	Poisoning campaign.
1890	Atlas Bear	Africa's only native bear hunted to extinction.
1905–52	Wolf sub-species	Japan, Newfoundland, Alaska, Mongolia, Southern Rockies, Cascade Mountains, British Columbia, Texas, Northern Canada. Hunted to extinction.
1935–79	Big Cat sub-species	Bali Tiger (1935), Barbary Lion (1942), Caspian Tiger (1970), Javan Tiger (1979). Hunted to extinction.
2011	Eastern Cougar (US); West African Black Rhinoceros	Human activity, loss of habitat.
2013	Formosan Clouded Leopard	Human activity, loss of habitat.

Past extinction events are known to have occurred over long periods of time, sometimes taking thousands of years. Estimates for what the human race has triggered and what we are now living through indicate that current rates of extinction are ten to a hundred times faster than any previous extinction event in Earth's history. As the studies so far show, in the second millennium there were eight recorded extinctions from the eleventh to the sixteenth centuries, four in the seventeenth, five in the eighteenth, sixteen in the nineteenth and forty-seven in the twentieth. As the pace of extinction accelerates, fourteen extinctions have already been reported as having occurred sixteen years into the third millennium. The rate of loss is becoming exponential: 'One in five species on Earth now faces extinction, and that will rise to 50% by the end of the century unless urgent action is taken. That is the stark view of the world's leading biologists, ecologists and economists'.[50]

The Nectar of Life

In space and time

Life on Earth wouldn't be possible without water: 'We (Allah) made every living thing from water' (21: 30). When astronomers search for planets outside of our solar system with prospects of harbouring life, they look for signs of water. No water, no life; and so far as we know, without water neither we nor any other life forms can exist. Attitudes to water vary according to where we live. Those of us who live in the lush tropics and temperate zones and near large lakes and flowing rivers are lackadaisical with the stuff. Those living in arid zones and desert fringes tend to be far more careful. But there are exceptions. Those with a surfeit of spending power can overturn the general rule, like those in the oil rich Middle East. The Gulf States rank amongst the highest per capita water consuming nations in the world.

Rain provides the illusion that there is an endless supply of water up in the sky, but we know that rain is part of the water cycle. It is finite and it has been recycled ever since it was created when the Earth's systems stabilized aeons ago. It keeps us and everything else on the planet alive, renewing, growing and cleansed. The water cycle is confirmation, if ever one was needed, of the interconnectedness of all things on the planet: 'We (Allah) drive(s) rain to the barren land' (32: 27). We live in a closed system. On average, human beings are about 70–80 per cent water. It is possible for water that was once in our bodies today to be quenching the thirst of another person living some distance away tomorrow. I may have been showering today with the water an Australian living in Sydney may have used to water his garden a few months ago. For this to happen, the sun will have done its work in drawing water as moisture from a garden into a cloud, it comes down again as rain, fed into a local river, carried out to sea and taken to the other side of the world by powerful currents, drawn up once again into the clouds, come down as rain into a local river, diverted to a local reservoir and then piped to where I live to enjoy my shower. Here's a thought: water connects us with our ancestors. It is recycled time and time again to be used by generation after generation. Water connects us with space and time.

Water is essential to maintain life and yet we are treating the Earth's oceans, freshwater systems, swamps and wetlands – all part of one system as we now know – with the same kind of disdain, indifference and disregard that we show the forests. Here are two examples, the first how we treat water as an agro-industrial resource and the second is the use of water resources as a weapon of war.

The Aral Sea was shared between the Central Asian states of Kazakhstan and Uzbekistan, both formerly part of the now defunct Soviet Union. It was one of the four largest lakes in the world and was estimated to occupy an area covering 26,300 square miles (68,000 square km) in the middle of the last century. The Republic of San Marino in Northern Italy would have comfortably fitted into this lake. It has now shrunk to 10 per cent of its original size.[51] In the 1960s, the old Soviet government initiated a programme to promote cotton growing in the arid regions of Central Asia by diverting the Amu Darya and Syr Darya rivers that fed the Aral Sea. Supplied by a network of canals that branched out from these rivers, the cotton bloomed and production targets were met. But the consequences of these events have been described as one of the planet's worst environmental disasters, destroying in addition a once-prosperous fishing industry and a way of life. 'We can now admit without any doubt that the Aral Sea crisis is the result of a large and brutal human impact, followed by the interaction between complex mechanisms present in nature.'[52]

When the Soviet Union broke up in the early 1990s it left a damning environmental legacy of which the Aral Sea debacle is but a part. This is no less than human aggression against the natural world on a grand scale and is yet another familiar example of what we are doing to the environment in the name of economic progress. We seem incapable of learning the lessons from this kind of debacle. The ecosystems of the Aral Sea region, including the river deltas that fed it, have been utterly destroyed and the entire region is left with a legacy of high salinity, toxic poisoning as a result of weapons testing, and poisonous dust storms. These are seen as the causes, taking a heavy toll on people's health resulting in 'unusually high'[53] fatality rates and high infant mortality.

The Iraqi Marshes lie in southern Iraq. Known to historians as Southern Mesopotamia, it is reputedly the seat of human civilization where agriculture and the first cities appeared. The marshes lie between the Tigris and the Euphrates rivers, mostly within Iraq with small areas in Kuwait and Iran. The direct descendants of the ancient Sumerians, who are today known as the Madan or Marsh Arabs, live there and can thus claim to be part of a civilization that is 5,000 years old.[54] The Sumerians dug irrigation canals and built dams for flood protection and in the process probably marked the beginnings of what we know today as the science of hydrology. Man has manipulated the environment as far as we care to remember but in modern times the nature of this intervention has been transformed both in terms of scale and complexity.

In 1951, Frank Haigh, a British engineer, prepared a report that proposed to drain the marshes by a series of sluices, embankments and canals in the Euphrates–Tigris river delta. The original purpose of the draining of the marshes was land reclamation for agriculture and oil exploration. But in 1991, Saddam Hussein used this device as a punishment against the Shiites in response to their rebellion against him after the First Gulf War. This triggered what was the almost complete destruction of the third largest wetland ecosystem in the world and by the mid-1990s the region had been reduced to a desiccated landscape and the once-thriving fishing industry was wiped out. The United Nations Environmental Programme called this action the worst environmental disaster of the last century, even taking into consideration the destruction wrought on the Kuwaiti oil fields by Saddam Hussein's retreating forces towards the end of the war.

The Iraqi Marshes once covered 7,700 square miles (20,000 square km) but by 2000 they had been reduced by 90 per cent of its original size. In the 1950s, there were over half-a-million Marsh Arabs living in this habitat, but by the mid-1990s there were only 20,000. The marshes were home to forty species of bird that numbered in their millions. It was also part of the migration route of millions of other birds that included herons, flamingos and pelicans as they travelled between Africa and Siberia. Since the overthrow of the Saddam Hussein regime in 2003, water flow to the marshes has been restored and the ecosystem has slowly begun to recover.

But there is more to come and this brings us face to face with present-day realities – the internecine conflict in the region notwithstanding. The Euphrates and Tigris rivers originate in Turkey, which plans to build a series of dams that will serve its energy requirements.[55] Turkey is no longer the 'sick man of Europe' and is now among the world's twenty largest economies. Turkey's Ilısu Dam, planned to deliver 2 per cent of its energy needs, is being built on the Tigris and will also irrigate 1.5 million hectares of land close to the Iraqi border. Syria is also siphoning off water from the Euphrates in large quantities. All this is set to reduce fresh water flow to the Iraqi Marshes and adversely impact the people relying on it for agriculture and fishing (see also chapter 3, 'The Progress Trap').

There are three issues here and they could be summarized as conflict, economics and the environment, the last occupying its usual place at the bottom of the list of priorities. Conflict is foreseen over increasingly scarce supplies of water as nations upstream take more for themselves, thus leaving less for nations downstream. Economics propelled by the growth model will tax scarce resources as never before and drive nation-states that should be trading with each other to compete for these very same resources, like water for example. The environment as always brings up the rear, such that by the time we realize that our actions have unforeseen and dire consequences it is nearly always too late.

Deep oceans and biodiversity

Seventy per cent of the Earth's surface is covered by the oceans. Its coral reefs are its rainforests and they teem with life, from minute plankton at the bottom of the food chain to giant whales, the largest animals that have ever lived. The biodiversity of the oceans is greater than that found on land and estimated to be between 50–80 per cent of the total. This rich marine biodiversity is suffering the same fate as its land-based cousins. Oceans play a major part in maintaining the CO_2 balance, but like terrestrial ecosystems they suffer when this balance is disrupted. Global warming is now creating this imbalance, as about 35 per cent of anthropogenic CO_2 dissolves in the sea, which is more than it can cope with. This is causing the seas to become more acidic and is one more reason for the bleaching of coral reefs, causing the marine life that depends on them to be threatened. Coral flourishes in optimum conditions in a symbiotic relationship with the algae that live in them and when these conditions are threatened by temperature changes and pollutants the coral turns completely white and dies off. At a local level there is a concern about poor fishermen threatening coral reefs with dynamite fishing, but even when this is hopefully dealt with the wider industrial-scale destruction of the coral reefs will continue. The Great Barrier Reef, described as Australia's natural wonder, 'is in mortal danger. Bleaching caused by climate change has killed almost a quarter of its coral this year (2016) and many scientists believe it could be too late for the rest.'[56] A survey of the reefs of the Maldive Islands in the Indian Ocean 'found bleaching on a devastating scale.'[57] The survey found 60 to 90 per cent bleaching in some sites.

What can we do about this? As usual in these matters the finger always points to those of us who think we just sit at home causing no harm to anything or anyone. These events may occur in ocean habitats far away from where we live but the interconnectedness of the natural world, and us to it, holds each one of us responsible for its welfare. And as The Nature Conservancy points out, 'You don't have to be a scientist to have a positive impact on coral reefs'.[58] Of the ten easy steps it proposes to protect coral reefs, six we can accomplish ourselves. They are conserving water, reducing pollution, putting a stop to the use of chemical fertilizer in our gardens and vegetable patches, disposing of waste sensibly and planting a tree. The tree connection should be obvious by now: the carbon the trees absorb would otherwise have found its way into the ocean.

Industrial fishing methods now threaten existing fish stocks and it is estimated that there are enough trawlers to fish three planets the size of the Earth. Factory ships circle the globe, fishing with nets that stretch for forty kilometres. They not only do immense damage, as they scoop up everything from the sea bed, but they also deprive local fishermen of their birthright. The catch is then cherry-picked for the most saleable items and the rest is thrown back into the sea, either dead or dying. Much of these dead discards are juvenile fish that cannot be sold and get dumped with little or no thought given to stock replenishment. Treating young fish like this leaves little scope for future fishing and future generations. Tuna and cod are so overfished that they are now reaching non-recovery levels.[59] According to the UN, 75 per cent of the world's fish stocks have been depleted to non-commercial levels. One of the consequences of this type of mindless predatory fishing is that it doesn't leave much for the small fishermen, who have relied for generations on local catches for their protein. The poor suffer once more at the grasping hands of the rich and powerful.

West African nations have some of the richest fishing grounds in the world; yet, their food security is under threat. European and Asian fishing fleets have moved into West African waters over the past thirty years after depleting their own fish stocks. Sub-Saharan Africa is now the only region on Earth where per capita fish consumption is actually falling, partly because foreign fishing fleets have removed so much fish.[60] The same is true for the Indian Ocean fisherman of East Africa, and in this instance when the fish ran out they turned to piracy.

One doesn't have to be a gardener to know that when a plant or weed is pulled out of the ground a ball of soil attached to its roots comes with it. Magnify this a few million times and one gets an idea of what happens to the soil when a huge area of forest is clear cut or torched. The soil at the base of the trees is not just disturbed but totally destabilized. When the rains arrive, as they always do, the trees are no longer there to protect the soil and a whole ecosystem – that has taken millennia to form – is washed away. This loosened soil then finds its way to the rivers and lakes and chokes them with nutrient-rich silt that kills the fish local people have relied on for their food supplies, and in extreme cases can divert rivers from their original course. Wooded slopes that are deforested can also be the cause of mudslides, which can bury whole villages and are now occurring with increasing frequency.

Mangroves are coastal forests found in the tropics and sub-tropics and are part of coastal ecosystems. They thrive in saline environments and are a protection against coastal erosion. They are also a protection against storms, and although no data is yet available it has been proposed that much of the damage caused by the Indian Ocean Tsunami in 2004 could have been prevented had the mangroves in the coastal zones remained intact. About 35 per cent to 40 per cent of mangroves have been lost in the past thirty years. Nothing is free from commercial exploitation: 28 per cent of mangroves, amounting to 1,344,000 acres (544,000 hectares) was lost to shrimp farming in this period.[61]

The Iraqi Marshes atrocity shows us the consequences of using nature as a weapon. Generically known as wetlands, marshes and swamps are natural ecosystems that teem with life. These wetlands are losing their biodiversity as are the forests. Wetlands host indigenous bird species and provide resting places for millions of migratory birds. For example, waterfowl migration from Alaska and Northern Canada to southern wintering grounds have been reduced by more than half from 145 million to 64 million mainly due to human encroachment.[62]

A Delayed Reaction

Global responses

To refer to our actions as human encroachment on the natural world is an indication that we have defined ourselves as the other. We cannot encroach on something of which we are already a part, it's more like tearing down a home one lives in. We have thought and talked our way out of the natural world, and having made nature 'the other' we are prepared to subvert it to suit our fancies. We call this 'progress' and 'development' and we have reduced the natural world to a mere resource that we plunder and pillage at will. And we persist in degrading the Earth in spite of increasing evidence that we are undermining our own existence. Seyyed Hossein Nasr describes this condition aptly: 'There is nearly total disequilibrium between modern man and nature as attested by nearly every expression of modern civilisation which seeks to offer a challenge to nature rather than to co-operate with it.'[63]

We now look for answers as the chickens come home to roost. Rachel Carson, the precursor to Theo Colborn, to whom I referred earlier, wrote *Silent Spring* in 1962.[64] That was more than fifty years ago and her book was a wakeup call that could not be ignored. She was the first to make the connection between pesticides, pollution and human health and was rightly credited with ushering in the modern conservation movement. The knock-on effect this triggered ultimately led to our flirtations with sustainability and this time span of fifty-five years has seen a revolution in environmental thinking. Or it could be argued that it was more of a panic reaction, which ultimately ended with the much lauded and much delayed Paris Agreement in 2015.

What emerged during this period, particularly at the top end of the institutional spectrum, are a proliferation of organizations, conferences, declarations and a plenitude of platitudes framed against a background of perpetual crisis. Despite the massive effort that has been put into reversing current trends, the responses never matched the gravity of the problems we are now facing. Structural changes didn't go deep enough and behavioural change at the individual level, where ultimately it really matters, is insignificant compared to the enormity of what we are now facing. A brief foray into the workings of the international system in this field will give us some idea of how, for the first time in human history, so much has been done by so many with so little to show for it.[65] One could say in mitigation that although much has been done, the strong tide of the problems caused by human 'progress' have more than cancelled out the gains made in the policy sector. Progress is the tide that has swept us into deeper currents.

With hindsight, these international endeavours could be seen as three separate but overlapping responses to resolve the issues that Rachel Carson uncovered: (1) conservation and biodiversity; (2) environment and development;

and (3) climate change. Although these categories are somewhat artificial they do give us a sense of the concerns that have been addressed globally, for example:

Conservation and Biodiversity

- UNESCO convened the Intergovernmental Conference for Rational Use and Conservation of the Biosphere in 1968.[66] Here there was collective recognition, probably for the first time, that natural resources were being used irresponsibly and irrationally. Early discussions on the ideas associated with ecologically sustainable development.
- The *Global 2000* Report was commissioned by US President Jimmy Carter (1977–81) to study 'the probable changes in the world's population, natural resources, and environment through the end of the century'. Released in 1980, the report recognized biodiversity for the first time as 'critical to the proper functioning of the planetary ecosystem'.[67]
- In 1982, the UN 'World Charter for Nature' added to the work done by *Global 2000* and adopted the principle that 'every form of life is unique, warranting respect regardless of its worth to man', and called for an understanding of our dependence on natural resources and the need to control our exploitation of them.[68]
- The Millennium Ecosystem Assessment (MA) was established to study 'the consequences of ecosystem change for human well-being.' Published in 2005, the same year the Kyoto Protocol entered into effect, it concluded that human actions are depleting Earth's natural capital and is putting such a strain on the environment that the ability of the planet's ecosystems to sustain future generations can no longer be taken for granted. At the same time, the assessment showed that with appropriate action it is possible to reverse the degradation of many ecosystem services over the next fifty years, but the changes in policy and practice required are substantial and not currently under way. Over 1,300 experts from ninety-five countries provided scientific evidence to the project.[69]

Environment and Development

- An international panel of experts produced the *Founex Report* in 1971. It identified industrially advanced countries' high levels of economic development and their large productive capacities as being responsible for the damage caused to the human environment, which was also threatening the whole world.[70]
- The International Institute for Environment and Development (IIED) was set up in the United Kingdom 'to seek ways for countries to make economic progress without destroying the environmental resource

base',[71] a contradiction of intentions if ever there was one. The flaws in this position are already obvious from the preceding discussion.

- The International Union for the Conservation of Nature (IUCN) released the *World Conservation Strategy* report in 1980. The section addressing Sustainable Development identified the main agents of habitat destruction as poverty, population pressure, social inequity and trading regimes. The report called for a new international development strategy to redress global inequities. This was guarded language and a shift in emphasis compared to the *Founex Report*. Lifestyles and wasteful unsustainable consumerism in developed countries were not addressed directly as matters of concern.[72]

- International Conference on Environment and Economics convened by the OECD in 1984 recognized the connection between the environment and economic development and concluded that they should be mutually reinforcing.[73]

- The Brundtland Commission produced *Our Common Future* in 1987 and added social and cultural dimensions to environment and economics. Brundtland was responsible for popularizing the idea of 'sustainable development'.[74]

- The Independent Commission on International Development Issues published *North–South: A Programme for Survival* (also known as the Brandt Report) in 1980, calling for a new economic relationship between North and South. In the little space it devoted to the environment it concluded that the strain on the global environment derived from the growth of both the industrial economies and populations. It called on all nations to 'cooperate more urgently in international management of the atmosphere and other global commons, and in the prevention of irreversible ecological damage.'[75]

- The UN Conference on Environment and Development (UNCED), popularly known as the Earth Summit, was held in Rio de Janeiro in 1992.[76] Agenda 21, 'a wide-ranging blueprint for action to achieve sustainable development worldwide', originated at this summit and it was described as 'a comprehensive plan of action to be taken globally, nationally and locally by organizations of the United Nations System, Governments, and Major Groups in every area in which human impacts on the environment.'[77]

- The UN's Millennium Development Goals were launched in 2000 at the largest-ever gathering of world leaders and acknowledged that, 'major gaps remain in reducing vulnerabilities for many developing countries, including least developed countries (LDCs), small island developing States (SIDS) and other low-income countries.'[78]

Climate Change

- Global warming first predicted at a scientific conference held in Villach, Austria in 1985.[79]
- Intergovernmental Panel on Climate Change (IPCC) established in 1988 by the World Meteorological Organization (WMO) and the United Nations Environment Programme (UNEP) to 'prepare, based on available scientific information, assessments on all aspects of climate change and its impacts, with a view of formulating realistic response strategies'.[80]
- The Kyoto Protocol, which committed national governments to reduce global warming, came into force in 2005 and was due to be renewed in 2012.[81]
- UN Framework Convention on Climate Change (UNFCCC) was adopted by 196 countries in Paris on December 2015. This agreement set a new legally binding framework for an internationally coordinated effort to tackle climate change. One of its major objectives was to attempt to limit global warming to 1.5 degrees Celsius.[82]

Abusing abundance

The human assault on the planet has been mindless, brutal and total. The relentless battering we have inflicted upon the environment in modern times in search of the good life has been deliberate and premeditated, illustrating the degradation of our relationship with the natural world. We have ravaged the Earth from the depths of the ocean all the way to the highest point in the stratosphere, and the scale, frequency and longevity of the types of events we have been discussing have their origins in recent history. Pre-industrial societies were biodegradable, and as they died off the forests regenerated and the sands covered their traces. We are abusing abundance, and it would seem that nothing we do is exempt in our pursuit of a consumer lifestyle that appears to take precedence over everything else. We are in a progress trap (see chapter 3, 'The Progress Trap').

The concern we need to demonstrate on behalf of the generations that will succeed us nowhere matches the magnitude of what we face. Stories abound of people standing on the shoreline watching incredulously as the sea rolled back before the 2004 Indian Ocean tsunami struck. As the sea girded itself and roared forward to the shoreline, it was too late to run. Climate change is of a different order in that it neither rolls backwards nor forwards: it builds up by stealth and strikes us at leisure. Unlike the tsunami and the shoreline, with climate change it is the whole Earth that will be affected – we have nowhere else to run – and public lethargy on climate change is alarming. Perhaps it is all too big to comprehend and people seek comfort in diversions, and there are plenty of those around.

The Earth will continue to be blasted and battered by humans for its rapidly declining non-renewable resources, and the consequences of this will continue to be unpredictable: weather, shrinking forests, loss of biodiversity, warming seas, the death of coral and depleting fish stocks, depleting stocks of fresh water, land, sea and air pollution, food scarcity, and the appearance of new diseases – and this is only the half of it. As indicated in the introductory paragraphs to this chapter, my intentions were to lay down as many of the facts as space would allow as regards the current state of planet Earth, environmentally speaking. The news is bad, and if this book is about anything I argue that the nature of change should be of such an order that people recover control of their own lives if we are to return the Earth to a semblance of balance. The globalized entities that run our lives are incapable of doing this. Collectively they lead the human community in the opposite direction, and the evidence is there for all to see.

I have been asked for good news and it is not that I have been reluctant to give any; it is not difficult to find individuals and organizations working at all levels confronting the issues I have been describing. For example, trees are being planted energetically by thousands of people all over the world. The Norwegian government is working with United Nations agencies to set up a faith-based global tropical rainforest protection programme. Given the enthusiasm with which people engage in these projects, I hesitate to say that any progress that is seen as being made in most of these endeavours is temporary and illusory. This is because the issues are systemic and they are related to the foundations that hold up modernity. I will attempt to give answers in the concluding chapters, but in the next two chapters I will endeavour to discover how we got into the hole we are in now, in the hope that it will provide us with some idea of how we can get out of it.

Corruption has appeared in the land and sea,
for that men's own hands have earned,
that He (Allah) may let them taste some part of that which they have done,
that perhaps they may return (30: 41).

Endnotes

1. Ralph Waldo Emerson, 'The Ministry of Nature', in Camille Helminski (ed.), *The Book of Nature: A Sourcebook of Spiritual Perspectives on Nature and the Environment* (Bristol: The Book Foundation, 2006), p. 84.
2. Hadith related by Muslim from Abu Saʿīd al-Khudrī.
3. Caspar A Hallmann, Martin Sorg, Eelke Jongejans, et al., 'More than 75 percent decline over 27 years in total flying insect biomass in protected areas' (*PLOS ONE*, 18 October 2017). Available at: http://journals.plos.org/plosone/article?id=10.1371/journal.pone.0185809 (accessed October 2017).
4. Philip J Landrigan, Richard Fuller, Nereus J R Acosta, et al., 'The *Lancet* Commission on pollution and health' (*The Lancet*, 19 October 2017). Available at: http://

www.thelancet.com/journals/lancet/article/PIIS0140-6736(17)32345-0/fulltext (accessed October 2017).

5. Fiona Harvey, 'Ocean acidification is deadly threat to marine life, finds eight-year study' (*The Guardian*, 23 October 2017). Available at: https://www.theguardian.com/environment/2017/oct/23/ocean-acidification-deadly-threat-to-marine-life-finds-eight-year-study (accessed October 2017).

6. 'Physical and human causes of climate change' (BBC Bitesize, undated). Available at: http://www.bbc.co.uk/education/guides/z3bbb9q/revision/6 (accessed on 3 May 2018).

7. Noel Castree, 'Has Planet Earth Entered New "Anthropocene" Epoch?' (Live Science, 30 August 2016). Available at: http://www.livescience.com/55942-has-planet-earth-entered-new-anthropocene-epoch.html (accessed on 3 May 2018).

8. In addition to carbon dioxide, greenhouse gases include water vapour, methane, nitrous oxide and ozone. They are essential to the stability of the earth's temperature and maintained an optimum mean level of around 15°C before the Industrial Revolution. Without GHGs the average surface temperature of the earth would be around -18°C. There has been a 40 per cent increase in atmospheric carbon dioxide since the Industrial Revolution, from around 1750, thus causing the global warming we are experiencing today. See, 'What are greenhouse gases?' (What's Your Impact, undated). Available at: https://whatsyourimpact.org/greenhouse-gases (accessed on 3 May 2018).

9. 'Villach Conference (Global Warming)' (what-when-how, undated). Available at: http://what-when-how.com/global-warming/villach-conference-global-warming/ (accessed on 3 May 2018).

10. For more information on the Intergovernmental Panel on Climate Change, see the official website at: http://www.ipcc.ch (accessed on 14 July 2018).

11. 'United Nations Framework Convention on Climate Change' (United Nations, 1994), Article 2, p. 9. Available at: http://unfccc.int/files/essential_background/background_publications_htmlpdf/application/pdf/conveng.pdf (accessed on 3 May 2018).

12. '[Kyoto Protocol] KP Introduction' (United Nations, undated). Available at: https://unfccc.int/process/the-kyoto-protocol (accessed on 3 May 2018).

13. 'Climate Change 2014: Synthesis Report Summary for Policymakers' (IPCC, 2014). Available at: https://www.ipcc.ch/pdf/assessment-report/ar5/syr/AR5_SYR_FINAL_SPM.pdf (accessed on 3 May 2018).

14. Emily E Adams, 'Eco-Economy Indicators: Carbon Emissions (Earth Policy Institute, 23 July 2013). Available at: http://www.earth-policy.org/indicators/C52 (accessed on 3 May 2018). I am grateful to Rutgers, The State University of New Jersey, for permission to use the accompanying graph.

15. 'The Paris Agreement' (United Nations, undated). Available at: http://unfccc.int/paris_agreement/items/9485.php (accessed on 3 May 2018).

16. Andrew Simms, '"A cat in hell's chance" – why we're losing the battle to keep global warming below 2C' (*The Guardian*, 19 January 2017). Available at: https://www.theguardian.com/environment/2017/jan/19/cat-in-hells-chance-why-losing-battle-keep-global-warming-2c-climate-change (accessed on 3 May 2018).

17. 'Climate Change 2014 *Synthesis Report*: Fifth Assessment Report' (IPCC, 2014), Section 2.1. Available at: http://ar5-syr.ipcc.ch/index.php (accessed on 3 May 2018).

18. 'Greenhouse gas concentrations surge to new record' (World Meteorological Organization, 30 October 2017). Available at: https://public.wmo.int/en/media/press-release/greenhouse-gas-concentrations-surge-new-record (accessed on 3 May 2018).

19. 'Cars produced this year' (Worldometers, undated). Available at: http://www.worldometers.info/cars/ (accessed on 3 May 2018).

20. Op cit.

21. 'Car choice impacts your green score and your cash' (Automobile Association, 11 January 2017). Available at: http://www.theaa.com/motoring_advice/car-buyers-guide/cbg_toxics.html (accessed on 3 May 2018).

22. Tech Metals Insider, 'How "Green" is Lithium?' (Kitco, 16 December 2014). Available at: http://www.kitco.com/ind/Albrecht/2014-12-16-How-Green-is-Lithium.html (accessed on 3 May 2018).

23. 'Chlorofluorocarbons (CFCs)' (Imperial College, London, undated) Available at: http://www.ch.ic.ac.uk/rzepa/mim/environmental/html/cfc.htm (accessed on 3 May 2018).

24. 'The Montreal Protocol on Substances that Deplete the Ozone Layer' (United Nations Environment Programme: Ozone Secretariat, undated). Available at: http://ozone.unep.org/en/treaties-and-decisions/montreal-protocol-substances-deplete-ozone-layer (accessed on 3 May 2018).

25. Paula Tejón Cardajal, 'HFCs and other F-gases: The Worst Greenhouse Gases You've Never Heard Of' (Greenpeace, undated). Available at: http://www.greenpeace.org/international/Global/international/planet-2/report/2009/5/HFCs-Fgases.pdf (accessed on 3 May 2018).

26. 'Kigali Amendment: Thirty Five Parties to the Montreal Protocol Have Ratified' (United Nations Environment Programme: Ozone Secretariat, undated). Available at: http://www.ozone.unep.org/en/focus (accessed on 3 May 2018).

27. Press Trust of India, 'Acid rains make life hard in 258 Chinese cities' (NDTV, 14 January 2011). Available at: http://www.ndtv.com/article/world/acid-rains-make-life-hard-in-258-chinese-cities-79213 (accessed on 3 May 2018).

28. 'Endocrine Disruptors' (National Institute of Environmental Health Sciences, undated). Available at: http://www.niehs.nih.gov/health/topics/agents/endocrine/ (accessed on 3 May 2018).

29. Carol F Kwiatkowski, Ashley L Bolden, Richard A Liroff, et al., 'Twenty-Five Years of Endocrine Disruption Science: Remembering Theo Colborn', *Environmental Health Perspectives*, 124/9 (September 2016), A151–A154. Available at: https://ehp.niehs.nih.gov/ehp746/ (accessed on 3 May 2018).

30. Theo Colborn, Dianne Dumanoski and John Peterson Myers, *Our Stolen Future* (London: Penguin, 1996).

31. Rosa Rozati, PP Reddy, P Reddanna and Rubina Mujtaba, 'Role of environmental estrogens in the deterioration of male factor fertility', *Fertility and Sterility*, 78/6 (December 2002), pp. 1187–1194. Available at: http://www.sciencedirect.com/science/article/pii/S0015028202043893 (accessed on 3 May 2018).

32. PCBs are a form of persistent organic pollutant. They were found to build up in the environment and cause harmful health effects. Their manufacture was stopped in the US in 1977.

33. Jean Carper, *100 Simple Things You Can Do To Prevent Alzheimer's and Age-related Memory Loss* (London: Vermilion, 2011), p. 107.

34. News Release, 'The cost of a polluted environment: 1.7 million child deaths a year, says WHO' (World Health Organization, 6 March 2017). Available at: http://www.who.int/mediacentre/news/releases/2017/pollution-child-death/en/ (accessed on 3 March 2018).

35. Alan J Jamieson, Tamas Malkocs, Stuart B Piertney, et al., 'Bioaccumulation of persistent organic pollutants in the deepest ocean fauna', Nature: Ecology & Evolution, 1, article no. 0051 (2017). Available at: http://www.nature.com/articles/s41559-016-0051 (accessed on 3 May 2018).

36. Claire Le Guern, 'When the Mermaids Cry: The Great Plastic Tide', (Coastal Care, March 2018). Available at: http://coastalcare.org/2009/11/plastic-pollution/ (accessed on 14 July 2018).

37. Op cit.

38. Vicki Pritchett, 'Deforestation in the United States' (Lovetoknow, undated). Available at: http://greenliving.lovetoknow.com/United_States_Deforestation (accessed on 5 June 2018).

39. *Khalīfah*, which I render as steward, is variously translated as successor, inheritor, heir, viceroy, vicegerent, or deputy.

40. For the official website of the Millennium Ecosystem Assessment see http://www.millenniumassessment.org (accessed on 14 July 2018).

41. The Arabic *najam* in this verse from the Qur'an lends itself to be translated both as planets or stars, thus connecting to the previous verse which refers to the sun and the moon. *Najam* mediates between the cosmic and the terrestrial.

42. Easter Island, situated in the Pacific Ocean, had a thriving civilization which collapsed in the seventeenth and eighteenth centuries. It is believed that extensive soil erosion caused by deforestation had contributed to its demise. See Jared Diamond, *Collapse: How Societies Choose to Fail or Succeed* (New York: Viking, 2005).

43. '51 Breathtaking Facts About Deforestation' (Conserve Energy Future, undated). Available at: http://www.conserve-energy-future.com/various-deforestation-facts.php (accessed on 3 May 2018).

44. 'Threats: Deforestation' (WWF, undated). Available at: https://www.worldwildlife.org/threats/deforestation (accessed on 3 May 2018).

45. The term 'ecosystem' is usually confused with 'environment'. While ecosystems are part of the environment it has a specific scientific meaning. There are many approaches to defining this but the two most useful are firstly the one that defines this in terms of their niche in the environment, and secondly the definition in terms of energy transfer. They are: (1) an ecosystem is a natural system formed by biological organisms interacting between themselves and their particular physical environment; and (2) a whole community of biological organisms living together linked by energy transfer.

46. Tony Juniper, *Saving Planet Earth: What is destroying the earth and what you can do to help* (London: Collins, 2007), p. 25.

47. Richard Leakey and Roger Lewin, *The Sixth Extinction: Biodiversity and Its Survival* (London: Weidenfeld and Nicolson, 1996), p. 245.

48. This data is culled from: Viviane Richter, 'The Big Five Mass Extinctions' (Cosmos, undated), Available at: https://cosmosmagazine.com/palaeontology/big-five-extinctions (accessed on 3 May 2018).

49. SL Pimm, GJ Russell, JL Gittleman and TM Brooks, 'The Future of Biodiversity', *Science*, 269 (1995), pp. 347–350. See also Gerardo Ceballos, Paul R Ehrlich and Rodolfo Dirzo, 'Biological annihilation via the ongoing sixth mass extinction signalled by vertebrate population losses and declines', *Proceedings of the National Academy of Sciences*, 114/30 (25 July 2017), pp. E6089–E6096.

50. Robin McKie, 'Biologists think 50% of species will be facing extinction by the end of the century' (*The Guardian*, 25 February 2017). Available at: https://www.theguardian.com/environment/2017/feb/25/half-all-species-extinct-end-century-vatican-conference (accessed on 3 May 2018).

51. Jeanne Kasperson, Roger Kasperson, and BL Turner, *The Aral Sea Basin: A Man-Made Environmental Catastrophe* (Boston: Kluwer Academic Publishers, 1995).

52. 'The Aral Sea Crisis' (Columbia University, undated). Available at: http://www.columbia.edu/~tmt2120/conclusion.htm (accessed on 3 May 2018).

53. 'Aral Sea: Impact on Social and Economic Sphere' (Web Archive, undated). Available at: http://web.archive.org/web/20090316062917/http://enrin.grida.no/aral/aralsea/english/arsea/arsea.htm (accessed November 2017).

54. Robert W Brown, 'Ancient Civilisation to 300 BC. Introduction: The Invention and Diffusion of Civilisation' (University of North Carolina, 2006).

55. Jim Muir, 'Iraq marshes face grave new threat' (BBC News, 24 February 2009). Available at: http://news.bbc.co.uk/1/hi/world/middle_east/7906512.stm (accessed on 3 May 2018).

56. Michael Slezak, 'The Great Barrier Reef: a catastrophe laid bare' (*The Guardian*, 7 June 2016). Available at: https://www.theguardian.com/environment/2016/jun/07/the-great-barrier-reef-a-catastrophe-laid-bare (accessed on 3 May 2018).

57. Gavin Haines, 'Paradise in peril: Maldives devastated by coral bleaching' (*The Telegraph*, 9 August 2016). Available at: http://www.telegraph.co.uk/travel/destinations/asia/maldives/articles/maldives-devastated-by-coral-bleaching-as-2016-shapes-up-to-be-hottest-on-record/ (accessed on 3 May 2018).

58. 'Coral Reefs of the Tropics: You Can Make a Difference' (The Nature Conservancy, undated). Available at: http://www.nature.org/ourinitiatives/habitats/coralreefs/ways-to-help-coral-reefs/ (accessed on 3 May 2018).

59. 'Why is overfishing a problem' (Overfishing - A global disaster, undated). Available at: http://overfishing.org/pages/why_is_overfishing_a_problem.php (accessed on 4 May 2018).

60. 'The ocean is our common resource: Exploitation off West Africa's coasts' (Greenpeace, undated) Available at: http://www.greenpeace.org/africa/en/campaigns/Defending-Our-Oceans-Hub/ (accessed on 4 May 2018).

61. Stuart Hamilton, 'Assessing the Role of Commercial Aquaculture in Displacing Mangrove Forest', *Bulletin of Marine Science*, 89/2 (Virginia Key, FL: University of Miami, Rosenstiel School of Marine and Atmospheric Sciences, April 2013), pp. 585–601.

62. Greg Yarrow, 'Wetland Ecology: Value and Conservation', Fact Sheet 32 (Clemson, SC: Clemson University Cooperative Extension, Forestry & Natural Resources Team, revised May 2009). Available at: http://www.dnr.sc.gov/wildlife/wetlands/ClemsonExtension-WetlandEcology.pdf (accessed on 4 May 2018).

63. Seyyed Hossein Nasr, *Man and Nature: The Spiritual Crisis of Modern Man* (London: Unwin Paperback, 1990).

64. Rachel Carson, *Silent Spring* (New York: Houghton Mifflin, 1962).

65. For a summary of the post-Carson initiatives that eventually emerged as Sustainable Development, see 'The Sustainable Development Timeline' (International Institute of Sustainable Development, 2010). Available at: http://www.iisd.org/pdf/2009/sd_timeline_2009.pdf (accessed on 4 May 2018).

66. 'Intergovernmental Conference of Experts on the Scientific Basis for Rational Use and Conservation of the Resources of the Biosphere' (Paris: UNESCO, 1968). Available at: http://unesdoc.unesco.org/images/0001/000172/017269eb.pdf (accessed on 4 May 2018).

67. Gerald O Barney, *Global 2000: Entering the Twenty-First Century* (Arlington, VA: Seven Locks Press, 1980). Available at: http://www.geraldbarney.com/Global_2000_Report/G2000-Eng-7Locks/G2000_Vol_One_7Locks.pdf (accessed on 4 May 2018).

68. 'World Charter for Nature' (UN General Assembly, 28 October 1982). Available at: http://www.un.org/documents/ga/res/37/a37r007.htm (accessed on 4 May 2018).

69. See the Millennium Ecosystem Assessment (http://www.millenniumassessment.org).

70. Miguel Ozorio de Almeida, *Environment and development: The Founex report on development and environment* (New York: Carnegie Endowment for International Peace, 1971). Available at: http://www.stakeholderforum.org/fileadmin/files/Earth%20Summit%202012new/Publications%20and%20Reports/founex%20report%201972.pdf (accessed on 4 May 2018).

71. The International Institute of Environment and Development. See http://www.iied.org/ (accessed November 2017).

72. IUCN, World Conservation Strategy: Living Resource Conservation for Sustainable Development (IUCN-UNEP.WWF, 1980). Available at: https://portals.iucn.org/library/sites/library/files/documents/WCS-004.pdf (accessed on 4 May 2018).

73. For a position on environment and economics, see Markku Lehtonen, 'The environmental-social interface of sustainable development: capabilities, social capital, institutions', *Ecological Economics*, 49/2 (2004), pp. 199–214. Available at: http://netedu.xauat.edu.cn/sykc/hjx/content/ckzl/2/2.pdf (accessed on 4 May 2018).

74. World Commission on Environment and Development, *Our Common Future* (Oxford: Oxford University Press, 1987). Available at: http://www.un-documents.net/our-common-future.pdf (accessed on 4 May 2018).

75. Independent Commission on International Development Issues, *North–South: A Programme for Survival* (London: Pan Books, 1980).

76. 'Earth Summit: UN Conference on Environment and Development (1992)' (United Nations, 1997). Available at: http://www.un.org/geninfo/bp/enviro.html (accessed on 4 May 2018).

77. 'Agenda 21: UNCED, 1992' (United Nations: Sustainable Development, undated). Available at: http://sustainabledevelopment.un.org/content/documents/Agenda21.pdf (accessed on 4 May 2018).

78. 'News on Millennium Development Goals' (United Nations: We can end poverty, undated). Available at: http://www.un.org/millenniumgoals/ (accessed on 4 May 2018).

79. 'Villach Conference (Global Warming)' (what-when-how, undated). Available at: http://what-when-how.com/global-warming/villach-conference-global-warming/ (accessed on 4 May 2018).

80. 'History' (IPCC, undated). Available at: https://www.ipcc.ch/organization/ organization_history.shtml (accessed on 4 May 2018).

81. UNFCCC, 'Kyoto Protocol'. Available at https://unfccc.int/resource/docs/convkp/ kpeng.pdf (accessed on 14 July 2018).

82. 'The Paris Agreement: Summary' (Climate Focus, December 2015). Available at: https://climatefocus.com/sites/default/files/20151228%20COP%2021%20briefing %20FIN.pdf (accessed on 14 July 2018).

3

PROSPERITY IN PERSPECTIVE

Happiness and the Good Life

Happiness and GDP

Happiness is a universal aspiration. It is the basis of everything we do, and what else do we live for but to attain a state of contentment? But its realization is another matter and much depends on ones station in life, that is who we are, where we are and what is happening around us. One could say that happiness reduces in proportion to the responsibilities we shoulder, but this needn't be the case as we sometimes refer to a person who bears his or her responsibilities 'lightly'. This conjures up an image of a burden, and one part of the human family that should have none of it – or the least of it – is its children. This is why their proper nurturing is so important, since life without burdens can be the happiest in one's lifetime. Happiness as an emotion can prevail in our relations with others, our spiritual inclinations and also our material circumstances. It is connected to states we describe as contentment and wellbeing. Here are examples of what three well known figures have to say about happiness.

Be happy for this moment. This moment is your life.
Omar Khayyam (eleventh/twelfth century Muslim poet, philosopher and scientist).

Happiness is not something readymade. It comes from your own actions.
Dalai Lama (living head of Tibetan Buddhism, now exiled in India).

Success is not the key to happiness. Happiness is the key to success. If you love what you are doing, you will be successful.
Albert Schweitzer (Franco-German Christian theologian, philosopher).[1]

One discerns a theme in these aphorisms and they seem to be saying that happiness is a state one creates for one's self. However, it is no exaggeration to say that nowadays this condition is almost exclusively related to a state of material wellbeing, shaped by a technology-driven consumer society that has raised our expectations considerably. That is, raised our contentment threshold. Unhappiness to a great extent stems from being made to want things today that didn't exist yesterday, and much of this has to do with the deliberate creation of discontent. If evidence is needed about this condition then all one has to do is to turn on the television and watch it for an hour and more than likely we will be bombarded by advertisements selling new products, ranging from cars to candy, suggesting that if they were in our possession we would be much the happier for it. The ubiquitous box we call television is deeply embedded in the global matrix, which has now become our gateway to happiness – and this includes all flat screens, from the wall to wall variety in cinemas to the mobile phones in the palms of our hands.

We take every opportunity to wish our friends, relatives and acquaintances all the happiness they can find, and these sentiments are usually associated with births, anniversaries, marriages and other notable events in our lives – as the greeting cards industry bear testimony. Wishing another good health and prosperity is also commonplace, while the once universal prayer to be endowed with many children is fading away, perhaps in the interests of reducing global birth rates. One can have good health and be unhappy and one can also be prosperous and miserable. Some take the view that this latter condition is preferable to being happy and poor, a state encapsulated in the phrase, 'I'd rather be rich and miserable than poor and happy'. But there are others who take a broader view and hold that the idea of prosperity, which supposedly influences happiness, covers a wider compass.

Captain Cook, on encountering the Australian aboriginal people during one of his voyages, wrote in his journal: 'These people may truly be said to be in the pure state of Nature, and may appear to some to be the most wretched upon Earth: but in reality they are far more happier than ... we Europeans'.[2] It would seem that by their own standards the people of the first nation of Australia considered themselves to be prosperous and consequently happy, but they were soon to become wretched. Inundated by migrants from the West, what had once been reputedly the oldest continuous culture on the planet was rapidly overwhelmed by a different set of values introduced by European immigration. It would seem that today there is no value differentiation, happiness is universally measured by the watches on our wrists, the mobile phones in our pockets and the cars in our drives.

In an attempt to unravel this confusion, the UK-based Legatum Institute has not only defined prosperity but has also produced an index, ranking 149 countries from the most to the least prosperous.[3] The Legatum method attempts to demonstrate that 'prosperity increases as GDP per capita increases' and suggests that prosperity is more than just the accumulation of material wealth; but it is also the joy of everyday life (happiness) and the prospect of an even better life in the future (this is a strictly secular expression and is not a reference to an afterlife); this applies as much to individuals as well as nations. Legatum claims to be an 'authoritative measure of human progress, offering a unique insight into how prosperity is forming and changing across the world'. It ranks the most and least prosperous countries and is divided into nine sub-indices, which go to show that prosperity is not the simple matter we thought it was. Alternatively one may take the other view, that there are people out there making life unnecessarily more complicated than it really need be.[4] Here are the sub-indices:

Economic Quality
Business Environment
Governance
Education
Health
Safety and Security
Personal Freedom
Social Capital
Environment

A glance at the rankings and measures for 2017 indicate that Norway heads the table. The rest of the top twenty positions are all held by countries in Europe, North America and Australia, with the exception of Singapore in seventeenth position, the only Asian nation to figure in the elite. The bottom of the table is conspicuous by the overload of African countries, and both Afghanistan and Yemen rank among the lowest; the unremitting conflicts that have plagued these countries in recent years provides some explanation for this. Legatum attempts to go beyond GDP, which measures wealth, and considers prosperity in a wider context, which includes aspects of wellbeing such as safety, security and personal freedom. Prosperity, the condition from which we supposedly derive our happiness, is not then a simple matter, and the next time we wish this on someone perhaps we need to contextualize our feelings. A friend may visit us from a country with restrictive laws on press freedom and perhaps we should be specific and wish them prosperity in the Personal Freedom sub-index. But the success or failure of these subsidiary indices depend on GDP, as they are all interlinked. It comes as no surprise then that the countries that are at the top of the Legatum table are all countries with high GDP rates usually associated with high per capita incomes, the good life and the ostensible consumer happiness that it brings. This is an envious position to be in, especially when looking at this from an African standpoint where low GDP is endemic.

But consider this from an environmental perspective. If the analysis in the previous chapter ('A Ravished Earth') is anything to go by the prospects are quite dire for planet Earth. In the past 250 years or so human numbers have grown from a mere half a billion to more than 7.6 billion today. As industries expanded and produced more, so populations expanded and consumed more, progressively diminishing the Earth's scarce resources. If, as population forecasts suggest, human numbers are set to increase to 10 billion by 2050, it then becomes imperative to ask how the planet is going to cope with this prospect, especially if we accept the Legatum analysis and its version of prosperity based on GDP.

Happiness and GNH

However, not all countries are represented in this index. There are two contrasting nations which have been left out of the reckoning germane to this discussion for reasons best known to Legatum. They are the Maldives (also known as the Maldive Islands) and Bhutan. The former is almost entirely Muslim and the latter Buddhist. The Maldives are an island state of twenty-six atolls lying south west of Sri Lanka. It is one of the most dispersed countries in the world, consisting of nearly 1,200 coral islands which if cobbled together would occupy a land mass equivalent to that occupied by the island of Malta in the Mediterranean. Knowledge of this unique island chain arises from the fact that it lies abreast in the Indian Ocean, like an obstacle course offering a challenge to the navigators of old who plied the trade routes between the Far East and Africa and all points in between. However, within less than twenty years of Vasco da Gama, the Portuguese explorer, rounding the Cape of Good Hope, the Maldives were forced into vassal status, thereby perhaps holding the dubious distinction of being the first country to the east of Africa to be colonized by the Europeans.[5] Maldivian defences were no match for the Portuguese gunboats and by 1512–13 they had succumbed to the invaders and since then have been subjected to the vagaries of the western imperial project. The Dutch overtook the Portuguese in the Maldives in the mid-seventeenth century, and this nation of atolls was finally wrested from the Dutch by the British by the end of the eighteenth century, and continued to be part of the imperial project until the twentieth.

The Bhutanese suffered no such travails. Their country lies snuggled in the bowels of some of the highest mountains in the world, which guaranteed its inaccessibility. Its average elevation is about 2,400 metres (8,000 feet) above sea level and it has a land area a little bigger than Denmark. It is also sometimes mistakenly identified as the mystical Shangri La, of the James Hilton novel *Lost Horizon*. But this is not to say that the Bhutanese went totally unscathed by European colonialism, nor were they averse to carrying on a bit of feuding themselves. The first Europeans who visited Bhutan, according to records, were two Portuguese Jesuit priests in the seventeenth century at a time Bhutan was warring with Tibet. The Jesuits presented the Bhutanese leader with firearms, but he refused to accept their offer of service in his war.[6] The Bhutanese invaded and occupied the kingdom of Cooch Behar to their south in the late eighteenth century but they were later ousted from this principality by a helping hand from the British East India Company. Bhutan consequently shrank in size due to incursions by the British and formal relations between the two countries were not established until early in the twentieth century.[7]

In spite of these localized difficulties Bhutan managed to remain obscure until the 1970s. It entered the United Nations in 1971 and a young 16-year-old king took the reins the following year, after the demise of his father. As he surveyed a tumultuous world from his tiny mountainous kingdom he coined the phrase 'gross national happiness', committing his country to its traditional spiritual values alongside economic development.[8] The following is a cursory look at this idea.

'Gross National Happiness (GNH) measures the quality of a country in more holistic way (*sic*) [than GDP] and believes that the beneficial development of human society takes place when material and spiritual development occurs side by side to complement and reinforce each other.'[9] In contrast to the nine sub-indices of the Legatum Prosperity Index, the GNH index contains the following nine domains:

Psychological wellbeing
Living standards
Good governance
Health
Education
Community vitality
Cultural diversity and resilience
Time use
Ecological diversity and resilience

This was an attempt to think and act holistically and go beyond the merely economic indicators which GDP represents. What the Happiness domains and the Legatum Index have in common are Standards of living (Economy), Good governance (Governance), Health and Education. Legatum recently added Environment, almost as an afterthought, and thus coincides with Ecological diversity and resilience in the GNH index. What the Happiness domains have, which Legatum doesn't, are Psychological wellbeing, Community vitality, Cultural diversity and resilience and Time use, none of which depend on GDP and increase the rating of the Bhutanese initiative.

Quantifying psychological wellbeing and community vitality have proved problematic to measure, owing to their subjective character. It took a 6–7 hour interview to ascertain if someone was happy or not. Nevertheless, this idea was endorsed by the UN, some nations like Singapore toyed with the idea in the shape of 'social reserves', and it created much excitement amongst researches and academics; there were think-tanks, commissions, research studies conferences and even a UN resolution. But Bhutan remains the only country to continue with the experiment and there is a note of unhappiness to the Happiness project: Acting contrary to its stated position on cultural diversity and resilience, the Bhutanese have made refugees of its Hindu population, who had made the country their home since the nineteenth century.[10]

There is also another serious sting in this Bhutanese tale. A drug called 'television' was introduced in 1999, which was soon to be followed by 'Bhutan's first crime wave – murder, fraud, drug offences'. A country that was immunized against external influences by its geography and culture over the ages suddenly found itself 'crash-landing in the 21st century'.[11]

The Maldivians for their part must have appeared to the European sailors as indolent lotus eaters; surely a sign of contentment. Life was easy, there was fish aplenty and food was not hard to come by. Maybe the Maldivians had their GNH and didn't know it? They don't have a recorded history, and this gap was partly filled by two Europeans who were shipwrecked on the islands. The first was the Frenchman François Pyrard, who was shipwrecked in 1602.[12] He lived in the Maldives for five years, became fluent in the local language and left a detailed record of the times. Harry Bell was a British archaeologist who suffered the same fate as Pyrard and was shipwrecked on the Maldives in 1879.[13] He became instrumental in providing the colonial Government of Ceylon (present day Sri Lanka) with a survey of the islands during the approach of the twentieth century. To Bell must go the credit for introducing the word atoll, derived from *atolu*, the original Maldivian, into the English vocabulary.

Persian and Arab traders are known to have visited in pre-Islamic times, and the islanders mainly have South Indian/Hindu and Sri Lankan/Buddhist roots. Islam arrived in the twelfth century. Insignificant they may have been, but this was where much of the cowrie shell, used as 'coinage' even up to the early part of the twentieth century, came from. Not to stretch a point, used across China and Africa the cowries must have given the Maldivians the status of world bankers, although they were not to know it. The Maldivians also produced coir rope, which came from coconut fibre, much in demand by sailors for its salt-resistant qualities. They also exported dried fish in great quantities. Today the Maldives are simply a tourist attraction, providing temporary respite and perhaps happiness to those who toil in the developed world.

Bhutan and the Maldives are as far apart as they can get in their geography, history, cultural traditions, civilization, religious affiliations and social mores, but today they are one in that they are victims of agendas set for them from elsewhere. Invaded by modernity, they have their problems like the rest of us. Minnows in a globalized world, the Bhutanese worry about forest denudation caused as a result of increased firewood collection and pollution caused by industrialization and urbanization, and the Maldivians by fear of being overwhelmed by the sea as the Earth warms up. The Maldivians have dedicated an entire atoll, called Thilafushi, in which they dump the waste resulting from the growth of tourism and the accompanying increase in local consumerism. They no doubt aspire to enter the Legatum league and crawl up its index, as other nations do, and thus contribute to the factors that spell their own destruction.

A prerequisite for happiness is contentment and there are as many states of contentment as there are people. Bhutan and the Maldives each had theirs, but today more than at any other time in human history events outside the control of the ordinary citizen, ostensibly connected with economic wellbeing, influence this condition. This is illustrated by the fact that millions of ordinary people through no fault of their own, including the Bhutanese and the Maldivians for sure, were adversely affected as a result of the global banking collapse of 2008. In a pithy analysis of economics in the last decade of the last century, the American economist John Kenneth Galbraith observed: 'Contentment sets aside that which, in the longer view, disturbs contentment; it holds firmly to the thought that the long run may never come.'[14] But will contentment ever come? The advertising hoardings create perpetual discontent, and rapidly evolving technology – like the ubiquitous smartphones, for instance – ensures a kind of short term pseudo-contentment.

Writing about technological society, Ernest Braun observes: 'Happiness depends on whether our expectations are fulfilled or disappointed, on whether reality is compatible with our dreams or bears no resemblance to them.'[15] The economist Richard Layard has made an attempt to objectify happiness and he asserts that happiness comes from both outside and from within: 'The secret is compassion towards oneself and others, and the principle of the Greatest Happiness is essentially the expression of that ideal.'[16] These views, expressed by both a scientist and an economist, are not entirely incompatible with each other in that the technologist creates the product and the economist the dreams. But the economist seems to want to take his ideas further and has launched Action for Happiness, which he describes as 'a mass movement for a happy society'. He asks, 'do we want a society that relies so heavily on self-interest rather than on commitment to the welfare of others?'[17] Surely this is the realm of religion. Are we then back to the future?

There are three 'buzzwords' I shall be looking at in the rest of this chapter that are the drivers of the modern world. They are not only embedded in the lexicon of economics but have now become so much part of the language of our daily lives that we take them for granted. They are 'progress', 'development' and 'growth' and need to be redefined if we are to save the planet from further degeneration and thus save ourselves.

Do not forbid the good things Allah has made lawful to you
And do not overstep the limits.
Allah does not love those who overstep the limits (5: 87).

The Progress Trap

Lifestyle change

For whole swathes of our history, our ancestors had no idea that they were progressing. Conceptually this was an idea that was superfluous to their struggles and their needs, but if the changes they had shaped over vast intervals of time had not taken root we, their descendants, would not have 'progressed' in the way we think we have. In prehistory, when our ancestors sought an accommodation with the natural world in their struggle for survival they would migrate from one habitat to a more favourable one. There was also another kind of movement, which involved a lifestyle change – transition from one stage of existence to another – that made their lives less harsh as we see it today. It was a movement from just being gatherers to becoming hunter-gatherers – from herbivores to omnivores.

In their new role as hunters they fashioned weapons and implements out of stone, heralding the onset of the Stone Age culture. This was also the origin of technology – no longer using sticks and stones in their original crude state but fashioning them to serve a variety of purposes. That was about three million years ago. As they migrated north and south from the equatorial regions to colder extremes they processed animal skins into clothing to keep warm. They were familiar with fire, as it would frequently occur in their habitats through the processes of spontaneous combustion: the volcanoes that spewed out molten fire into the atmosphere and the lightning that set tinder alight. This caused them problems, but they succeeded in taming this force about a million years ago. This was perhaps the beginnings of the camp fire; a huddle of people keeping warm around a bright blaze.

Our ancestors learnt not only to communicate their experiences from one generation to the next but 'by developing cultures transmissible through speech'[18] they set themselves apart from other sentient beings. The agricultural revolution is estimated to have appeared about 12,000 to 10,000 years ago, after the glaciers retreated for the last time. It was not a revolution measured by time because it will have taken thousands of years of trial and error for certain communities in certain favoured locations to give up their hunting habits and become farmers. It is in this sense of a clear change in lifestyle that it can be called a revolution, and Homo sapiens seems not to have looked back since. From reliance on the hunt they became reliant on the seasons stabilized by post-glacial climatic changes, and conditions were appearing that made the growth of civilizations possible. They moved from caves to built shelters; from isolated communities to villages, to towns, to cities; from water in puddles to reservoirs and irrigation canals; from footpaths and hunting trails to paved roads; from food that lasted a day or two at most to grain stores that lasted a season; from animal skins to silks and high fashion.

With dire predictions of extreme climate change already upon us, we now wonder where we are heading. Thanks to our ancestors the generational transmission of knowledge has been developed to the extent that we now know our prehistory with a considerable degree of accuracy. The detail is in the zone of academic research, but the broad thrust is agreed upon and this allows for the discussion of a basic idea such as this. We learn from the past that all things come to an end, even civilizations; even the highly sophisticated global civilization we now inhabit, make no mistake. In a trenchant analysis of progress, the philosopher John Gray observes that Enlightenment thinkers 'with few exceptions' were neo-Christian 'missionaries of a new gospel more fantastical than anything in the creed they imagined they had abandoned. Their belief in progress was only the Christian doctrine of providence emptied of transcendence and mystery.'[19] Gray further opines that the secular 'religious impulse' has sought 'salvation through politics' which is now 'decidedly shaky':

> The grandiose political projects [Marxism, neo-Liberalism] of the twentieth century may have ended in tragedy or farce, but most cling to the hope that science can succeed where politics has failed: humanity can build a world better than any that has existed in the past. They believe this not from real conviction but from fear of the void that looms if the hope of a better future is given up. Belief in progress is the Prozac of the thinking classes.[20]

Like climate change deniers today, the Homo sapiens of bygone ages were in the habit of denying their own demise at the highest point of the blossoming of their civilizations. At this point specializations are highly developed and solutions tend to appear when it is too late, a phenomenon analysts describe as 'overshoot'.[21] And even if we know the solutions, self-interest gets in the way. If proof of this all-pervasive human condition is needed, witness the behaviour of US President Donald Trump. On taking office in January 2017, one of his earliest actions was to take his country out of the climate Agreement concluded in Paris in 2015. This Agreement was the culmination of a painfully negotiated process that had taken over ten years to complete, conducted by the international community in the full knowledge that any delay would be disastrous for us all. Is progress then a trap? In this we have much to learn from the past, it is a linear process and as there is a beginning there is inevitably an end.

Agriculture had arrived. A life-changing revolution, it has taken its time and paved the way for the growth of civilizations. Hunting and gathering in small groups was left behind and toiling on the land has become the norm. This required living in close proximity to the land where the seeds had been sown, which propelled the growth of village communities, then towns and then cities. As populations congregated into small spaces it triggered the growth of hierarchy, power, a priesthood and a ruling elite. It also meant war, a signal that civilization had arrived. Organized killing replaced inter-clan skirmishes. These transitions of what some might call progress, first took place in what is known today as the Fertile Crescent, an inverted crescent stretching from the Nile delta, today's Israel, Lebanon, northern Syria, southern Turkey, Iraq and western Iran. What is thought to be the first civilization was developed by the Sumer people in lower Mesopotamia, in the area mainly enclosed by the Euphrates and Tigris rivers. They were settling the area by the sixth millennium BCE,[22] developed a high civilisation by the fourth that had collapsed by the second. In this period the Sumerians created an administration, invented writing, which left us the Epic of Gilgamesh to savour the period and crop records to give us an understanding of their rise and fall. They built temples, monuments and Eridu, the first city in the world. They perfected the use of the plough, invented the wheel and pioneered the art of glass making. All these inventions, which we take so much for granted today, took place in the fourth millennium BCE.

At first glance, a map where the Sumer civilization rose and fell suggests an ideal location for farming and the growth of communities. Here we experience population growth, the first sign of the consequences of adequate food supplies. However, while it may have been adequate for a modest expansion of agriculture, it was not suitable for the continued support of a civilization growing in numbers and sophistication. For the early settlers on the river banks it looked promising, and the establishment of settlements further away from the rivers, supported by irrigation canals, was an organic extension of these first settlements. The idea caught on and crop surpluses helped build the first human civilization, but the Sumerians had no idea they were walking into a trap. We can describe their predicament in these terms today, with the benefit of hindsight, but unlike us they had no past to learn from. They were pioneers and what brought them down were the difficult conditions in the location itself. Farming between two rivers may have sounded like a good prospect but the climate was harsh; 'the summers were long and hot and the winters harsh and cold'.[23]

Water was difficult to access away from the rivers, rainfall was scarce and the flooding from melting snows that originated in the distant mountains in eastern Turkey and north west Iran came at the wrong time of the year. Motivation must have really been high to build a civilization out of these conditions, but build they did. What they did not realize was that the substance that was to give them hope was also to spell their end: Water. The terrain between the two rivers was not suitable for the intense, sustained agriculture of the kind that was practiced, and the nectar of life itself applied in copious quantities in inappropriate locations caused massive salinization. As the soil became progressively more saline there was a corresponding decrease in crop yield, and we know this from the records the Sumerians kept: 'The desert in which Ur and Uruk (reputedly the earliest cities to be built) stand is a desert of their own making'.[24] What has been described today as the overshoot phenomenon raised its head many millennia before our time, and it is something we should learn from (see also chapter 2, 'The Nectar of Life').

Remarkably similar events played themselves out in the region covering southern Mexico and Central America much later in the human story. Settlements in this region date back to the fifth millennium BCE, and their evolution into civilizations followed a pattern not a great deal different to the experience of their Sumerian cousins across space and time. Farming was being practised in the second millennium BCE, and the pre-Mayan Olmecs had by the first millennium BCE developed writing, the zero, astronomy and a calendar. How they collapsed as a civilization is not exactly known, and environmental factors and the collapse of the local ecology are offered as possible causes. But a great deal more is known of the Mayans who overlapped the Olmecs, eventually succeeding them in first millennium BCE. By the end of the first millennium CE,[25] however, the Mayan civilization had all but disappeared. Although the Ecology and environment of Mesoamerica was so utterly different to that of Sumer, human behaviour was much the same, with greed and the lust for power raising their ugly heads. Farming produces surpluses, which produces settlements, which produces population growth, which produces hierarchies. As in Sumer, internecine warfare was rife. Mayan kings outdid each other in building grandiose structures and short term considerations took priority. As pressure on the environment grew through the increase in population and the growth of urban centres, the hills became denuded of their forest cover, inevitably resulting in hillside erosion. The soil gradually lost its productive capacity due to nutrient loss as a result of over farming and, it is believed, human induced drought.[26] The people who produced the long count calendar, looking forward in vast periods of time, couldn't see what was coming just a few generations ahead of them.

Once Vasco da Gama rounded the South African cape in 1497, the race for the land grab east of Africa by the European powers was on. This was parallel to Christopher Columbus' feat on the Atlantic in 1492. By the eighteenth century, Portuguese, Dutch, British, American, French and Spanish sailors were scouring the vast empty spaces of the Southern Pacific in search of a mythical southern land mass, first postulated by Aristotle in the fourth century BCE. But what they discovered was something quite the opposite. They came upon a small semi-barren island just a little bigger than the Isle of Wight (380 square kilometres) and lined up on the horizon were a phalanx of huge stone megaliths. The Dutch sailors who were the first Europeans to come upon this island sighted it on Easter Sunday in 1722. They decided to call it Easter Island, and so it has remained on the maps ever since (it was annexed by Chile in 1888 and is known in Spanish as Isla de Pascua). What they encountered was 'a society in a primitive state with about 3000 people living in squalid reed huts or caves, engaged in almost perpetual warfare and resorting to cannibalism in a desperate attempt to supplement the meagre food resources available on the island.'[27] As it turned out, what these Dutch sailors had discovered was the remnants of a long-dead civilization. When succeeding anthropologists, archaeologists, paleo-botanists, carbon daters, historians, linguists, fossil hunters and all and sundry had done their work the following story emerges.

It is estimated that the first Polynesians arrived at what was then an uninhabited island about 500 CE. To accomplish this feat they needed to have travelled across anything between 2,500 and 3,000 km (approximately 1,500 and 2,000 miles) of open ocean. They did so in fragile catamarans, and needless to say without maps or other navigational aids. Their arrival at Rapa Nui (the native name for the island) is a wonder in itself, and a glance on a map at the vast empty spaces of the Pacific Ocean will tell one why. Many previous such expeditions would have been lost in this emptiness. The island was mainly wooded, with an estimated twenty-one species of large trees, multiple plant species and populated by land and sea birds. There were no mammals and what water there was gathered in the craters of extinct volcanoes when it rained. These Polynesian sailors may well have been the very first humans to set foot on the island, and with no streams or water courses to boast of they were not to know that the ecosystem they had inherited was fragile, to say the least. But thrive they did and the boat load of people who landed in the fifth century had grown to an estimated peak of 7,000 (some estimates put it at 15,000) by the middle of the sixteenth century. However, this was followed by a drastic decline and when the Dutch sailors arrived in 1722 they found a rump of people numbering less than 3,000.

What caused this sudden and rapid collapse? The short answer that is offered is that the Rapa Nui had denuded the island of its tree cover and hunted the birds to extinction. The trees were cut down to make clearings for agriculture, for building living spaces, boats for fishing, fashioning other domestic essentials and for use in the transport of the statues they were carving out of volcanic rock. On arrival, these islanders developed a sophisticated micro-civilization based on ancestor worship. The only visible sign of this civilization today are the massive stone statues called *moai*. Tree trunks were used to provide the rails to transport the *moai*, which weighed anything up to 80 tons and measured 10 metres (34 feet) in length, to their platforms. To further exacerbate matters, the production of *moai* increased with intensified clan rivalry, resulting in increased logging. Sustaining these activities resulted in the last standing tree on the island being felled. The ecosystem of Rapa Nui, which was in steady decline, ultimately collapsed – aided and abetted by the rats that were introduced to the island by the settlers. The rats bred fast and furious, had no predators and gnawed at everything they could get their sharp teeth into. This was perhaps one of the earliest lessons of the consequences of introducing an alien species in to an ecosystem to which it did not belong.

When the European sailors arrived they found the remnants of a people who were trying to survive in a habitat they themselves had destroyed. However, these people were not about to see the end of their travails. After the Dutch, British and Spanish had done their exploring in the eighteenth century, burgeoning modernity and globalization were raising their unpredictable heads in the nineteenth. Progress had arrived, and by the middle of the nineteenth century whalers had introduced tuberculosis. Rapa Nui was raided by slave traders from Peru in 1862, and when they were forced to repatriate the people they had kidnapped, the returnees carried smallpox, which had devastating consequences for the island.[28] These two diseases almost wiped out the Rapa Nuians. The first Christian missionary arrived in 1862, and by 1870 Jean-Baptiste Dutrou-Bornier, a French criminal, had taken virtual control of the island with the intention of turning it into a sheep farm. By this time the number of locals were reduced to fewer than two hundred. Dutrou-Bornier was subsequently killed and the island was taken over by Alexander Salmon Jr, the son of an English Jewish merchant, and a local princess. Salmon sold his holdings to the Chilean government in 1878, who annexed the island the following year. The island was subsequently rented to the Scottish owned Williamson–Balfour Company, who ran it as a sheep farm up to 1953. The company enclosed the local people in the Hanga Roa Township inside a boundary wall; a virtual prison for more than fifty years, a reflection of the Gaza strip in Palestine today. By 1966, the Rapa Nuians were fully fledged Chilean citizens. They are now part of the Indigenous rights movement, and the descendants of the original settlers who came ashore on their catamarans 1,500 years ago are now claiming the island back for themselves.

Progress has a price

We have looked at three examples of civilizational collapse in three different time–space conditions. Although the Sumerians, the Mayans and the Rapa Nuins couldn't be more different in their outlook and cultures, they all collapsed for remarkably similar reasons. Examples are rife in the history of human progress, as civilizations rise, decline and fall they stack themselves up, one on top of another, sometimes building indiscriminately on the experiences of the others. In doing so, they ignored the harsh lessons of those underlying fault lines that had destroyed their predecessors and built on the external manifestations of success. Our world today looks to be grander, more powerful, more knowledgeable and more prosperous than previous civilizations, enjoying an unheard of degree of technological sophistication, when in reality it has no more stability than the same old house of cards (see chapter 1). Ponting observes that, 'it was not until the end of the seventeenth century that the continuing increase in scientific knowledge and the steady advance of technology ... began to convince some thinkers that history might be a chronicle of progress rather than decay'. It came to be accepted by European intellectuals 'that history was a series of irreversible changes in only one direction – continual improvement' and marked in the eighteenth century 'by a wave of optimism about the future and the inevitability of progress in every field'.[29] This, however, was a particularly Eurocentric outlook, at a time when the riches of the world were opening up for them due to intrepid sailors and advanced technology: Their gunboats proved to be an irresistible force. The optimism of today is based on a particular mode of thinking that is already outdated. In a little over five hundred years after the voyages of Christopher Columbus and Vasco da Gama, planet Earth is now on the verge of severe multiple ecosystems collapse.

The first human who dug a hole in the soil in some bygone age and planted a seed also planted the idea that the environment can be manipulated to human advantage. We haven't looked back since, and there were no PR machines, intellectuals or scientists around to laud this seminal event. As we have seen, what followed was a pattern: adequate food supplies, growth in human numbers, the flowering of a civilization and then collapse as the land was denuded of its bounty – if in the meantime it was not otherwise destroyed by a natural calamity or an act of war. The difference today is that what was once local has now become global. Let's take note of the following ten issues, all of which we have seen before and are playing themselves out again today:

1) Population growth: There are over 7.6 billion of us today and set to reach around 9–10 billion by 2050.

2) Food shortages: In spite of the apparent plenty and waste in the so-called developed world, millions go hungry in other parts. With continuing population growth the forecasts of shortages are alarming.[30]

3) Environmental degradation: Ecological collapse in various parts of the world has already been set in train (see chapter 2, 'A Ravished Earth').

4) Extravagance: This takes the form of conspicuous consumption; a characteristic of the vainglorious human, now globalized.

5) Concentration of power: We needn't go further than the UN Security Council, where five nations hold a veto over all others. Then there is the nuclear arms club, the powerful banks that can make or break economies – as we have seen in their collapse in 2008 – and the ubiquitous multinational corporations, some of whom have greater budgets than some nation states.

6) Competition: The Rapa Nuians sculptured the *moai*, the Sumerians the Ziggurats and the Mayans their pyramids to outdo their rivals. We build mega cities, rapid transport systems, tall buildings reaching for the skies, huge dams that store vast lakes of water, weapons of mass destruction that can tear our planet apart and much, much more.

7) Conflict: War and internecine conflict are endemic.

8) Disease: To give just one example, we have seen what Smallpox did to the Mesoamericans as a result of the Spanish invasion. With the world shrinking thanks to air travel, the odds are in favour of a lethal global pandemic.

9) Denial: At one level there is an alarming lack of awareness about environmental issues, where avoiding the truth and all unpleasant news is part of the problem. At another, not an inconsiderable number of people actively deny the prevalence of human-induced climate change.

10) Reluctance to change: People are loath to give up lifestyles they believe they have worked hard to attain.

All this sounds rather pessimistic, but what good news is there for people on a sinking ship? Lifeboats is the obvious answer, but the analogy ends there as one cannot jump off the planet as one jumps ship. The answer lies in the last two points above, and uniquely the good news is created by each one of us. You don't wait for good news, you create it yourself. We'll discuss this in the final chapter of this book, but it is as well to be aware that the trap of progress lies in wait as we stumble our way through the tangled undergrowth towards prosperity.

> It is He (Allah) who produces gardens, both cultivated and wild,
> and palm trees and crops of diverse kinds, and olives and pomegranates,
> alike yet different.
> Eat of their fruits when they bear fruit and pay their due on the day of their harvest.
> And do not be wasteful for He (Allah) does not love the wasters (6: 141).

Development and Delusions

Corralling the world

When President Truman of the United States of America gave his inaugural speech in January 1949, he announced his regime's intentions to 'embark on a bold new program for making the benefits of [American] scientific advances and industrial progress available for the improvement and growth of underdeveloped areas'. In helping people to 'realize their aspirations for a better life' he offered his country's 'imponderable resources in technical knowledge which [were] growing and [were] inexhaustible' to achieve these ends.[31] The background to this was the rewriting of the global political map in the aftermath of the Second World War, which saw the decline of the imperial powers like Great Britain and France and the emergence of the USA and the Soviet Union as the two new super powers that controlled the destiny of the world. The Cold War between the capitalists and the communists in the aftermath of the war had only just begun, and before long the two super powers were building up nuclear arsenals, to all appearances to defend themselves, should the other be foolish enough to strike first.

Like the invention of the 'cold war', the idea of the 'nuclear deterrent' entered our lexicons. Each side championed the aspirations of what later became known as the 'third world', and what ostensibly protected the human race from Armageddon was the doctrine of Mutually Assured Destruction, which aptly lends itself to the acronym 'MAD'. So, while Native Americans were languishing in their reservations and black Americans were virtual third-class citizens, the American president was reaching out to what he described as the 'under developed' world. Whilst it would be churlish to deny an element of altruism in President Truman's intentions, his approach – in addition to gaining him valuable friends – also opened up the world's resources for the growth of

American big business, an increase in the prosperity of the American people and a widening of the wealth gap between themselves and the so-called under developed people.

The new world order was taking shape as the wreckage of the Second World War was being cleared away. The World Bank and the International Monetary Fund (IMF) were created in 1944 at the Bretton Wood conference held in New Hampshire in the United States. These institutions are now part of the United Nations system and are located in Washington DC. There were forty-four participating nations and the purpose of the gathering was to put some order into the international monetary and financial system, which had all but collapsed as the war was ending. It had evolved under the imperial mantle since the fifteenth century, and to allow this to collapse would have ensured the collapse of capitalism itself, just as communism was raising its head in the East. There was a second objective to Bretton Woods, and that was to maintain a degree of control over this new financial regime in the light of the collapse of the Gold Standard, which had proved to be so difficult to sustain prior to the war. International trade was to be freed by opening up global markets and there was no room for economic nationalism (this idea is now in disarray as a result of the actions of President Trump), although protectionism continued to flourish in the advanced countries. Significantly, 'There was an almost inevitable lack of concern for the interests of developing countries'[32] at that time. India, which was still three years away from independence, was represented by a delegation led by a British official.

Bretton Woods paved the way for economic imperialism to fill the gap left by the demise of the old imperialism. A new global map was taking shape and labels like 'first world', 'second world', 'third world', 'north', 'south', 'advanced nations', 'emerging nations', 'industrialized nations', 'industrializing nations', 'developed world' and 'underdeveloped world' were gaining common currency.[33] These were euphemisms for rich and poor nations and built into these definitions was the assumption that the latter needed to be modernized and were going to be assisted in catching up with the former. This scenario conjured up tantalizing prospects for the old colonial countries, who were themselves emerging into an uncertain world as fledging nations. And now the Americans were championing their cause. The noose of progress was tightening as the new nations were seduced into this trap (see chapter 1, 'A Lethal Cocktail'). As the Cold War intensified between the capitalist and communist blocs (first and second worlds) a movement emerged amongst the newly independent countries which sought an identity for themselves as neutral nations. This gave way to the Third World Movement, which saw its birth in Bandung, Indonesia in 1955. It was led by charismatic post-colonial leaders like Sukarno (Indonesia), Jawaharlal Nehru (India), Gamal Abdel Nasser (Egypt) and Josip Broz Tito (Yugoslavia). The part that is of interest to us is in the Final Communiqué of the conference, which underscored the need for *developing countries* (author's italics) to loosen their economic dependence on the leading industrialized nations by providing technical assistance to one another through the exchange of experts and technical assistance for developmental projects, as well as the exchange of technological know-how and the establishment of regional training and research institutes.[34]

This was the beginning of the Non-Aligned Movement, which saw itself as being neutral in the East–West Cold War. Participants also agreed to promote Afro-Asian economic and cultural cooperation and to oppose colonialism in all its forms. Of the twenty-five countries that participated in this conference all but six of them were ex-colonial countries and even these had not been free of western interference in their affairs. The third-worlders, as they described themselves, were going to help each other, but they themselves needed help in order to achieve what they had set themselves up to do.

Colonized psyches

It would seem, however, that the Truman Road Map was not working as it should since the wealth gap between the developed and the underdeveloped countries continued to grow. The Independent Commission on International Development, popularly referred to as the Brandt Report,[35] published its findings in 1980 and found imbalances in technical expertise, manufacturing capacities and terms of trade as being some of the main problems that caused the wealth gap to widen. In other words, the advantage the developed nations had built up over the underdeveloped during the period of colonial expansion continued to grow. Brandt urged a transfer of resources, which was largely ignored, and the wealth gap has continued to widen to this day.

The movement towards solidarity between the ex-colonial nations was given a further boost when the South Commission was formed in 1987, 'following years of informal discussion between intellectual and political leaders from the South'.[36] One of the prime movers behind this initiative was Mahathir Mohamad, then Prime Minister of Malaysia and now at the helm again, who persuaded Julius Nyerere, the retired President of Tanzania, to chair the Commission. An interesting partnership if ever there was one, the former a democrat in the capitalist mould and the latter a socialist with Marxist–Leninist leanings and close ties to China, which at that time was functioning within the Maoist mould. Apart from their common colonial past, as part of the British Empire, they had in common an aspiration on behalf of their people to catch up with the living standards of the people who once ruled them. The well-intentioned report contained some of the following objectives of interest to our discussion. It observed that the 'benefits of prosperity and progress' have passed them by and expressed a 'desire to escape from poverty and underdevelopment' being experienced by the South.[37] The Report saw development as a process that enabled people to 'realize their potential, build self-confidence and lead lives of dignity and fulfilment'. It also saw development as a means of giving political independence its true meaning and saw in '[economic] growth, a movement essentially springing from within the society that is developing'.[38] These aspirations in a sense unwittingly endorsed colonialism, because implicit in what they were proposing was a devaluation of their own ancestral cultures by seeking to emulate a model of progress that suited the intentions of their erstwhile colonial masters. They were locked into the paradigm of the other by the imperial project.

In wanting to match the North in the manner it had developed, the leaders of the South were speaking with colonized psyches. The world they had inherited was severed from the world of their fathers and they were seduced into an order which was not of their making. They had swallowed and were being swallowed by modernity, and the kind of progress they were looking for came at the cost of sacrificing much of their cultural legacies – and of course destroying their environment. It needs to be noted that every single one of the post-war, post-colonial leaders, from Nehru to Nyerere, were educated in the western mould; even Nasr, who never left Egypt for his education, was known to have had a particular interest in the French philosopher, Voltaire. The imperial project had done such a thorough job in destroying indigenous cultures that leaders of the emerging world were thinking more like their ex-colonial masters, in turn reflected in the languages they used to communicate with each other: English, French, and Spanish, which suited the North very well. 'To be sure "development" had many effects, but one of the most insidious was the dissolution of cultures which were not built around a frenzy of accumulation'.[39] Yet the South persisted, and this manifested itself in a grand objective: a desire to emulate the economic success of their old masters. What was not foreseen was the cultural suicide this would entail. The job was so thorough that the erstwhile politicians who took over the mantle of leadership during decolonization in the mid-twentieth century presided over the seamless entrenchment of the imperial legacy, as the Bandung initiative signifies. Any opposition to this was branded as fundamentalism, tribalism or backward looking and the mission school 'educated' classes were in control. The battleground was economics and the issues ideological: Capitalism versus Marxism. It would seem that we have all inherited the Bandung legacy seamlessly, with the proviso that Marxism lost out to Capitalism. The 'frenzy of accumulation' continues unabated to this day: We all go shopping.

What also escaped notice was that as it grew and developed over the past 500 years, northern might had a global outreach and was built on plundered resources. They took what they wanted, when they wanted, from every nook and cranny in the world at their pleasure and leisure. This dubious disadvantage weighs heavily against the economically aspiring South, which still continues to be the primary source of raw material for Northern development. And since Bandung the air has been rife in third world gatherings with slogans of development, mutual assistance and self-help. 'The South was thus precipitated into a transformation which had long been going on in the North: the gradual subordination of ever more aspects of social life under the rule of the economy'.[40] We are now partners in the assault on the natural world on a massive scale in our pursuit of development.

Ideology

The Oxford Dictionary defines ideology as a 'system of ideas and ideals, especially one which forms the basis of economic or political theory and policy', and this is precisely what development has become: a universal ideology often crossing the boundary of a religious faith. In examining the ethics of development, Denis Goulet identifies a set of three values which in his view are desired by all societies: optimum life-sustenance, esteem and freedom. 'They refer to fundamental human needs capable of finding expression in all cultural matrices and at all times.'[41] Noble Laureate Amartya Sen widens the scope of this subject by arguing that real development 'requires the removal of major sources of unfreedom: poverty as well as tyranny, poor economic opportunities as well as systematic social deprivation, neglect of public facilities as well as intolerance or overactivity of repressive states'. He argues that there is no necessary correlation between increasing opulence and elementary freedoms in the contemporary world and 'Sometimes the lack of substantive freedoms relates directly to economic poverty'.[42] For Sen perhaps Singapore remains an underdeveloped state, as the sophisticated Singaporeans have accepted a measure of repression to achieve one of the highest standards of living in Asia.[43] In addition to Singapore, Hong Kong, Brunei and the oil rich countries of the Middle East all do well in the per capita incomes league and they are all countries that suppress civil liberties in varying degrees. This poses the interesting question whether freedom is a necessary concomitant to leading the good life.

Ideologies need institutions to make them work, and in addition to the World Bank and IMF, which takes care of the resource side of the equation, it is the United Nations Development Programme (UNDP) that sets the political tone. The UNDP was created in 1965 with the amalgamation of its technical assistance programme and its dedicated special fund. It took on a new meaning in 1990 with the introduction of the Human Development Index (HDI), which was formulated by Pakistani economist Mahbub ul Haq and a team of development economists with the support of Amartya Sen. An Inequality Adjusted Human Development Index (IHDI), seen as a more sophisticated evaluation of development, was introduced in 2010. It does not come as a surprise in the 2016 Report that northern countries figure by and large at the top of the table and the sub-Saharan African countries at the bottom.[44] This result has remarkable similarities to the Legatum Prosperity index discussed earlier. More than sixty-five years after Truman made his call, the northern countries continue to hold the top positions while the sub-Saharan African countries still languish at the bottom, with the wealth gap widening with unnerving regularity. This is not for want of trying.

Hope and sustainable development

The idea of sustainable development was popularized by the World Commission on Environment and Development report, *Our Common Future*, popularly known as the Brundtland Report, which was published in 1987.[45] This report was an attempt at refocusing the nature of economic development following the discovery that the natural world was being denuded by human activity. Brundtland defined sustainable development as:

> ... development that meets the needs of the present without compromising the ability of future generations to meet their own needs. It contains within it two key concepts:
>
> • The concept of 'needs', in particular the essential needs of the world's poor, to which overriding priority should be given: and
> • The idea of limitations imposed by the state of technology and social organization on the environment's ability to meet present and future needs.

Nobody would disagree with the first of these propositions, as it is about meeting the needs of people with emphasis given to the neediest. But the second idea appears to be missing something. It does not acknowledge that further limitations may be imposed by the carrying capacity of the planet – beyond those of technology and social organization. There is no reference to the over-consuming developed world and no indication as to how, in a finite world, resources could be shared more equitably. There is a contradiction here in that this approach does in fact compromise 'the ability of future generations to meet their own needs.' Daily and Ehrlich wrote in 1992:

> Given current technologies, levels of consumption, and socioeconomic organization, has ingenuity made today's population sustainable? The answer to this question is clearly no, by a simple standard. The current population of 5.5 billion[46] is being maintained only through the exhaustion and dispersion of a one-time inheritance of natural capital ... including topsoil, groundwater, and biodiversity. The rapid depletion of these essential resources, coupled with a worldwide degradation of land ... and atmospheric quality ... indicate that the human enterprise has not only exceeded its current social carrying capacity, but it is actually reducing future potential biophysical carrying capacities by depleting essential natural capital stocks.[47]

In another approach to sustainable development, the historian Fred Spier opines that its 'definition is a little vague, given the fact that no one can know with any degree of certainty what future generations will need to meet their needs. Yet if they will need resources that are similar to the ones we are using today, this definition presents a major challenge to humanity.'[48] Spier then proceeds to quote from the 1975 foreword by James Fletcher, then NASA administrator, to the summary report on the Apollo moon expedition thus: 'Husbanding the planet's finite resources, developing its energy supplies, feeding its billions, protecting its environment, and shackling its weapons are some of these problems.'[49]

Sustainability is about managing depleting finite resources, energy supplies, food security, protecting the environment and dealing with conflict and population growth, which passed the 7.6 billion mark in 2017 and continues its alarming upward trajectory. In the midst of the wrangling that resulted in the search to find a successor to the Kyoto Climate Protocol, the United Nations produced the Millennium Development Goals (MDGs).[50] These eight goals ranged from eradicating poverty, to reducing child mortality, through to developing a global partnership for development, and were agreed upon at a United Nations summit in 2000 to be achieved by 2015. The results were mixed, and to take one example the MDG bureaucracy claims to have met with limited success in its poverty eradication programme in that it purportedly halved the proportion of people living on less than $1 a day by 2008. In the estimates of the World Bank this result was achieved mainly because of the performance of China and India.[51] These two countries taken together have more than a quarter of the world's growing population and they also boast very high growth rates, considered by economists to be unsustainable and by environmentalists to be disastrous.

The General Assembly of the United Nations duly adopted a post-2015 development agenda, 'Transforming our world: The 2030 Agenda for Sustainable Development', in September 2015. This was in anticipation of the Paris Climate Summit (COP 21) that was to follow in December of that year. Throwing caution to the winds, the first sentence in the preamble reads: 'This Agenda is a plan of action for people, planet and prosperity'.[52] Prosperity is not defined, but this does recognize the links between economic development and global warming. There are seventeen goals and, 'Each goal has specific targets to be achieved over the next 15 years'.[53]

In the concluding paragraph of the previous section in this chapter I said that in order to make progress in this matter good news needs to be created by each one of us, and the UN has been helpful in this regard by producing lists of things we could be doing in our space and time under the heading 'The lazy person's guide to saving the world':[54]

- Things you can do from your couch.
- Things you can do at home.
- Things you can do outside your house.
- Things you can do at work.

Here is an opportunity to be part of the solution. There is much to do and there are more than a few things in these lists that each one of us can be doing.

Thinking globally and acting locally is the way to proceed and in the meantime it would be as well to have some idea of the big picture.

Sustainable development is a massive exercise in global social engineering based on a model of development that had evolved in the North under totally different circumstances. An army of technocrats, economists and academics are attempting to foster a counterfeit egalitarianism against impossible odds on populations that are still striving to maintain a semblance of diversity in their cultures and modes of being. Although environmental concerns appear as one of the sustainability priorities, the emphasis given to it leaves much to be desired. Nor has it been sufficiently stressed that much of everyday human activity, however small, has an effect on the environment and the aggregated sustainability objectives will sooner or later negatively impact on the planet. This is not an argument against the laudable sustainable development objectives, but if at one and the same time the rich are not exhorted to tighten their belts to enable the poor to loosen theirs a little, there isn't going to be much left for 'future generations to meet their needs'. It is said that the road to hell is paved with good intentions.

While the sustainable development initiative focuses on the lower rungs of the economic order, those at the top race away with massive bank balances. The wealth gap between the poorest and the richest in China and India, two so-called emerging countries, have widened considerably. The *South China Morning Post* reported that 'Income inequality worsened for the first time in five years, with the top 1 per cent owning a third of the country's total wealth',[55] and in India *The Hindu* reported that 'India's 1 per centers – its super-rich ... holds close to half of the country's total wealth.'[56] The *Guardian* published a similar account in its reporting of Credit Suisse's global wealth report: 'The globe's richest 1% own half the world's wealth ... highlighting the growing gap between the super-rich and everyone else.'[57] What this demonstrates is that there are two worlds, light years apart from each other in the way they see economics and wealth creation. One represented by agencies who are conservationists in outlook and the other represented by big business, which is expansionist.

A cautionary tale

Here's a cautionary tale that might help clear some of the cobwebs surrounding development. My organization, IFEES,[58] has developed an environmental teaching tool based on the Qur'an called *Qur'an, Creation and Conservation*,[59] which we deliver in various parts of the world. As a response to the tsunami that hit Aceh and the Sumatran coast in December 2004, we were there delivering this workshop to develop a wider sense of environmental responsibility amongst the local people. The particular event that comes to mind is a conversation I had with a young local man during the course of a field trip that took us to Batang Gadis National Park located in the Mandailing region of South Sumatra. We had a sub-agenda, to look for the Durian fruit that grew in the forest. Ours was a largish group which included a professor from the University of North Sumatra, located in Medan on the East coast of Sumatra. To the uninitiated, this fruit could grow to the size of a football, has a thick prickly outer skin, and an aroma – which can in polite terms only be described as excruciatingly pungent – that can travel some distance. We sniffed our way to a small hamlet deep in the forest and managed to acquire a few Durian fruits. But how to cut this fruit open and share it? It requires some skill and special knives. As we were pondering over this a young man came to our rescue. He offered the verandah of his hut for shelter and also supplied the tools with which to cut open the fruit. With the salaams and greetings over the conversation drifted to the usual small talk and I asked our host (let's call him Ali), who seemed reasonably well informed, what he was doing. The following conversation then ensued (Ali spoke in Indonesian which was translated for me by the professor from Medan):

Ali: 'I am going to the University in Medan.' (This surprised us all.)

Me: 'What will you be studying?'

A: 'Forestry.' (More surprises.)

M: 'But you grew up in this forest and you know all about it!'

A: 'I went to the local school in Panyabungan. I need to know more about the forest.'

M: 'But your father protected the forest and he had no university degree. And so did his father and his father and so on. Couldn't you learn from the old ways and the old wisdom?'

A: 'I can, but I need to have more scientific knowledge and there is also illegal logging.'

M: 'But you can't stop illegal logging with science or a university degree. Illegal logging is about greed and irresponsibility. Will the University teach you how to deal with this?'

A: 'No. But as a graduate I will have more authority and I will encourage the development of this area. If people are better off they will not need to work for the illegal loggers.'

M: 'But your father and the village elders had authority and there was no illegal logging in their time. But won't development make more demands on the forest?'

A: 'Yes, but past generations were poor. I want a better standard of life for me and my family, when I have one, and contribute to the development of my country.'

M: 'That's not a bad thing. But if you live in Medan where all the jobs are you will live far from the forest which you want to save. And to visit this area it will mean a forty minute flight plus a three hour journey by road, or alternatively a twelve hour journey by road, which will mean that you will be contributing to climate change with a very heavy carbon footprint.'

A: 'I will work locally and live in Panyabungan which is the local township.'

M: 'But Panyabungan may have already reached the extent of its development. It is crowded and its population is already encroaching on the forest and contributing to its degradation. Where will you live?'

A: 'I'll get my own house built.'

M: 'But the wood for this will come from the forests your fathers managed to save and it may come from illegal logging.'

Ali's reply to this was inaudible. We parted after promises to meet again and continue this conversation on another occasion.

Feedback loop

Empathizing with Ali's thought processes will in a sense point us in a direction that could conceivably cause us all discomfort, because his route to prosperity is now the universally accepted one. We are all Ali – what he wants is what we all want and the good or bad news is what we make up as we pursue life's struggles. The issues are systemic and we are in a feedback loop: education, careers, consumerism, taxing the environment, more education, more consumerism, resulting in an aggregation of environmental collapse. How to escape this is something I will be examining in the final chapter of this book.

Writing in 1992, Wolfgang Sachs observed: 'The last forty years can be called the age of development. This epoch is coming to an end. The time is ripe to write its obituary.' He then adds, 'Four decades later, governments and citizens alike still have their eyes fixed on this light flashing just as far away as ever: every effort and every sacrifice is justified in reaching the goal but the light keeps on receding into the dark.'[60] And for more than twenty years since development has been the proverbial carrot that many 'developing countries' have not been able to reach. Much to the encouragement of people who don't see development as defined for us with rose tinted glasses, a more recent alternative view appears to be emerging from deep within the bowels of the academic world itself. Henrietta Moore, in a lecture delivered to an Oxford audience and broadcast by the BBC World Service on 7 March 2015, declared roundly that development was an outmoded concept. She argued that this sixty year experiment has failed and technical fixes and infinite growth were not the answers, and that 'the fatal flaw of "development" is that it is a concept invented by the global North and imposed on the global South.'[61]

All this must sound rather gloomy, but optimism – being an essential quality of human nature – never ceases to raise its head. I was invited to a global forum in Oslo to attend a symposium on tropical rainforests organized by the United Nations Development Program and the Norwegian government. This event, convened as the Interfaith Rainforest Initiative, issued a Statement of Participants on 21 June 2017 addressed to leaders in government and business, to leaders and followers of our spiritual and religious traditions, and to the wider human family. Here are its first few introductory lines:

The Earth's rainforests are an irreplaceable gift.

They support boundless biodiversity, a balanced climate, and the cultures and communities of indigenous peoples who live in them. They generate cooling air and rains that water the Earth. They are spectacular, and vital to all life.

And they are at grave risk.

We, people of many faiths and spiritualities, gathered in Oslo to hear the cry of Earth's rainforests, their flora and fauna, and the people who live in them. We are Indigenous, Christian, Muslim, Jewish, Hindu, Buddhist, Daoist, joined by scientists who share with us, and open for us, a deeper appreciation of the miracle of the forests. We are from 21 countries – from Amazonia, the forests of Indonesia, the Congo Basin, Meso-America and South and Southeast Asia and the Pacific Islands, and from the US, Canada, Europe, and China. While from many places, we recognize that we are one human family, that we share one Earth.[62]

But spoilers still abound. Just about two months later, *The Guardian* carried the following headline:

'Brazil abolishes huge Amazon reserve in "biggest attack" in 50 years'

The Brazilian president Michel Temer has abolished an Amazonian reserve the size of Denmark, prompting concerns of an influx of mineral companies, road-builders and workers into the species-rich forest.[63]

Sorry, but that's development!

The Fantasy of Growth

The 'Holy Grail'

Everything that comes into existence grows, matures, flourishes, perishes and finally decays. Then, if the conditions are right there is renewal. These are unalterable universal laws; growth is the flip-side of development. The maturing and flourishing dimensions are the developmental aspects of growth. In another sense the terms 'development' and 'growth' are interchangeable, but in economics they have specific meanings, like in the sentence: 'Development is dependent on growth.' But we can have development without growth. That is, improving our state of wellbeing without economic growth; without the accumulation of material possessions that, in their increasing manufacture and use contribute to a degraded planet. We need to differentiate between quality and quantity and ask if possessions can improve the quality of life. Up to a point, perhaps, when they can relieve us of our daily drudgery. But beyond that we succumb to greed, euphemistically described today as consumerism; a state by which we measure our self-worth through the things we possess, capitulating to advertising techniques designed to leave us addicted to wanting more stuff. This is the consequence of denying ourselves the secret we call contentment, an essential ingredient for happiness. The extreme example of the desire to possess is the millions the rich spend on works of art. Somehow the possession and display of these artefacts give their owners a sense of power and importance, and perhaps happiness.

Growth has become the holy grail of the nation state in these times; it is the driver that provides the raison d'être of the modern world. One never ceases to hear about economic growth, as if this was the universal remedy that would deliver to the human race the good life it seeks. Politicians are sold on the growth agenda because this is how they can make their contribution to improving the living standards of their people. This is, after all, what they were elected to do and if they succeed they remain in power. The project is to enhance their appeal to the electorate, and this is the underlying reason why it has been difficult to reach agreement on climate change discussions. Reducing carbon footprints means curbing growth rates, and who is going to be the first to do that? This is why the United States did not ratify the Kyoto protocol in 2005, and Trump provides us with a textbook example of how this is done: tell people the lie they want to hear and scupper international agreements ostensibly in the national interest.

International agencies like the World Trade Organization (WTO) constantly promote growth. As Roberto Azevêdo,[64] the Director-General of WTO, declares in his concerns for the environment, the fact cannot be ignored that the principle concern of his organization is economic growth.[65] Economists always promote growth because it is part of their belief system, and in their reckoning the environment is relegated to second place. As we move into unexplored territory a new jargon appears: the 'ecology of investment', and in this idea investors sensitive to climate change look for growth opportunities in the development and application of renewable energy. The media, which with a few exceptions is part of big business, supports the growth agenda to the hilt. One only has to look at the financial pages in the newspapers. The following headline appeared in the *Telegraph* of 30 September 2014: 'IMF: Infrastructure spending spree last chance to revive growth – International Monetary Fund describes public infrastructure spending as "one of the few remaining policy levers available to support growth"'.[66] This is an organization in which its chief officer has declared her grave concerns regarding climate change. To bring this up to date, the latest IMF assessment published in October 2017 makes this observation in its Executive Summary: 'For 2018, the upward revision (a reference to its foregoing analysis) mainly reflects an expectation that the authorities will maintain a sufficiently expansionary policy mix ... to meet their target of doubling real GDP between 2010 and 2020.'[67] The full title of this report is *World Economic Outlook, October 2017: Seeking Sustainable Growth (Short-Term Recovery, Long-Term Challenges)*. It would be pertinent to ask here if sustainable growth is the same as sustainable development? It quite clearly is not and this raises another question. Is the IMF paying heed to the idea of sustainability that UNDP is taking so much pain to advocate? These are mixed messages coming from reputable international agencies and only confirms my two-world theory: conservationist versus expansionist.

If we draw a line starting with Adam Smith, the father of modern economics (late eighteenth century) and work our way through John Stuart Mill, economic theorist and moral philosopher (early nineteenth century), and Alfred Marshall, whose Principles of Economics was taught in universities till the 1920s, we discover that economic growth per se didn't receive any serious mention until the middle of the twentieth century, when economic historian Walt Whitman Rostow made a serious study of this subject in its own right. He saw the rate of growth as a 'function of changes in two enormously complex variables': Output, that is scale and productivity of the workforce; and, capital, which he describes as 'land, other natural resources as well as scientific, technical and organizational knowledge'. He also describes growth as 'the art of interrelating economic, social and political factors over time.'[68] Growth today is expressed as the amount of goods and services produced per head of the population over a period of time.

The late 1940s and the early 1950s were a period of great hope. Seventy years ago, the American atom bombs dropped on Hiroshima and Nagasaki had done their work and they provided the illusion that war was outmoded. Post-war reconstruction was reaching a climax, full employment was an aspiration treasured by all politicians, and the welfare state was being put together as a fulfilment of the hopes of war-weary people. Environmental issues were on the horizon, but few people saw them coming. For the first time ever an elected government was able to promise its people a good life by quoting a magic figure. This feat was achieved by RA Butler, Chancellor of the Exchequer (Finance Minister) of the British government, in 1953: 'He told the British people that their standard of life, with a 3 per cent growth rate, would double in 25 years.'[69] Over sixty years later this idea has now taken a grip on the world as a measure of economic progress. The 'UK Economic Outlook, March 2018' projects that 'households will spend over 30% of their budget on housing and utilities by 2030, up from around 27% in 2017.'[70] The culture is to view this as good news, and why not? Promises of the good life lie behind such statements. At the other end of the growth spectrum are China, who have at times achieved double-digit growth rates, and India who follow close behind. And again, why not? The people of these populous nations have a right to a higher standard of living and to be able to go shopping, illusory though this may be.

The economic growth agenda that the international order has designed for itself has taxed ecosystems to such an extent that the process of global decay is now well underway. The potent mix of factors, not least of which is climate change, have been simmering for a while and are now coming to the boil. The planet will recover in its own time, once we have done our mischief and are gone, but in the meantime we are in trouble. I have been around for nearly four-score years and ten and I have yet to see signs that the earth is growing. As far as I am aware, and people keep reminding me, if not stopped in time only cancer and the money supply can keep on growing forever. These two strange bedfellows also have another thing in common, in that if allowed to grow unchecked they will destroy their hosts. However, the growth of the former can be stopped by surgical incision but the latter, which in effect feeds the growth agenda, will need a different kind of surgery to stop it from killing its obese host. There is no Plimsoll line – a point at which we can all see that we are overloading the planet by these very processes.[71] Here is a perspective on growth from the New Economics Foundation:

> The most recent data on human use of biocapacity sends a number of unfortunate signals for believers in the possibility of unrestrained growth. Our global ecological footprint is growing, further overshooting what the biosphere can provide and absorb, and in the process, like two trains heading in opposite directions, we appear to be actually shrinking the available biocapacity on which we depend.[72]

Planetary overload

Some agencies have been busy measuring planetary overload and overshoot. On 13 August 2015, I received an email circular from the World Wide Fund for Nature (WWF) telling me that, 'Today is Earth Overshoot Day which means that in less than eight months, we have used more natural resources than the planet is able to produce in a single year. For the rest of 2015 we are living on resources taken from future generations'. By 1980 we exceeded the carrying capacity of the planet[73] and the Global Footprint Network[74] tells us that we use the equivalent of 1.7 planets to provide the resources we use and absorb our waste. Acknowledging that the developed world consumes a great deal more than the underdeveloped world we have, globally speaking, overshot the carrying capacity of the planet. This manifests itself in climate change, which we are all desperately trying to reverse. We can add to this land, sea and air pollution, rampant deforestation, species extinction, loss of biodiversity, collapsing fish stocks, acidifying oceans and many other effects which we are increasingly becoming familiar to us.[75]

At the same time there is a population explosion, there is also an explosion of consumption. It is not just the fact that we eat into the resources of a finite planet, thus denying future generations of its gifts, but we are doing it in a way which is mathematically described as exponential.[76] Linear growth is shown by an increase by a specific amount in a given period, but exponential growth is expressed as a percentage which is compounded. We see this factor at work in population growth; it is a doubling process, although the rate of doubling has currently eased off. Another element of exponential growth is that the period between each doubling shrinks. For example, it is estimated that there were a billion people in 1800. It rose to two billion in 1927, 127 years later. In 1974 the world population had doubled to four billion, just forty-seven years later, and currently (2018, forty-four years later) it stands at 7.6 billion. In the next chapter I will discuss a special case of exponential growth, in the hope that it will provide some answers to the dilemmas we now face.

Estimates suggest that if current population and consumption trends continue we are going to need the equivalent of two Earths to support us by the 2030s. On the other hand, if we begin a gradual cutback on our lifestyle patterns now it will take us until the 2050s to reach equilibrium between what the planet can give us and what we take from it.[77] There is, however, another special tribe in the human family who by all accounts live on another planet. The *Wall Street Journal*, whose readership largely consists of members of this tribe, carried a headline on 21 August 2015 that to its readers must have sounded like alarming news: 'Growth Worries Send Stocks Sharply Lower', and the report continued, 'Growth worries rattled stock markets from China to Germany to the U.S.'.[78] The worries referred to here were about potential shrinkages in industrial growth, drop in share values, economic downturn and financial turmoil. It would seem that some people, in whose hands much of the future of the planet depend, are oblivious to the consequences of their actions.

As we have seen, there are many pitfalls in our pursuit of universal prosperity and what the modern world has contrived to do is to mobilize Earth's entire population towards an unmitigated assault on the natural world, on whose declining wealth our desires rest. Aspirations are built on the generous use of buzzwords such as 'prosperity', 'progress', 'development' and 'growth', which unless redefined should be dumped altogether in the interests of saving the planet from further degradation. These terms have in various ways been used as economic indicators and, in spite of some advances, what this shows, if anything at all, is the prevalence of a de facto economic colonialism. The so-called Third World leaders that emerged after decolonization in the middle of the last century resembled their ex-colonial masters both in their attitudes and aspirations, having been suitably seduced by the black arts of the capitalist model. Marxism provided an alternative world view but in its application the environment suffered just as grievously. We are all now sucked into a process we now describe as globalization. De-growth, or negative growth, is now mustering a growing number of adherents, but the declining consumer demand that will result from this is seen as resulting in a spiral of recession.[79] If we huff and we puff we can blow the house of cards down. It seems obvious to me that a permanent global recession would save the planet for our children and this might be an area worth exploring.

In his *Prosperity Without Growth*, Tim Jackson quotes from *Redefining Prosperity*, published in 2003 by the UK Sustainable Development Commission, 'which challenged governments to "fundamentally to rethink the dominance of economic growth as the driving force in the modern political economy, and to be far more rigorous in distinguishing between the kind of economic growth that is compatible with the transition to a genuinely sustainable society and the kind that absolutely isn't".'[80] The penultimate word here lies with Jackson: 'The starting point for all this lies in a vision of prosperity as the ability to flourish as human beings – within the ecological limits of a finite planet'.[81] As we have seen we have far exceeded the planet's ecological limits and there are no signs of a return to some kind of sanity, sustainable development notwithstanding.

I would express my position in a simple equation:

$$\text{Progress} + \text{Development} = \text{Pollution} + \text{Destruction}$$
$$P + D = P + D$$

Endnotes

1. These quotes were culled from the online resource, BrainyQuote (https://www.brainyquote.com).

2. Brian W Richardson, *Longitude and Empire: How Captain Cook's Voyages Changed the World* (Toronto: UBC Press, 2005), p. 82.

3. The Legatum Institute is an international think-tank and educational charity focused on promoting prosperity. The index defines prosperity as a combination of wealth and wellbeing. See http://www.prosperity.com/#!/ (accessed on 15 July 2018).

4. Legatum Institute, *The Legatum Prosperity Index, 2017* (Legatum Institute, 2017). Available at: http://prosperitysite.s3-accelerate.amazonaws.com/3515/1187/1128/Legatum_Prosperity_Index_2017.pdf (accessed on 4 May 2018).

5. Bailey W Diffie and George D Winius, *Foundations of the Portuguese Empire, 1415–1850: Europe and the World in the Age of Expansion*, volume 1 (Minneapolis: University of Minnesota Press, 1977), p. 177.

6. Trent Pomplun, *Jesuit on the Roof of the World: Ippolito Desiden's Mission to Eighteenth-Century Tibet* (Oxford: Oxford University Press, 2009), p. 49.

7. Andrea Matles Savada (ed.), *Bhutan: A Country Study* (Washington: GPO for the Library of Congress, 1991). See 'British Intrusion, 1772–1907'. Available at: http://countrystudies.us/bhutan/9.htm (accessed on 15 July 2018).

8. '2015 GNH Survey Report' (Centre for Bhutan Studies & GNH, undated). Available at: http://www.grossnationalhappiness.com/ (accessed on 4 May 2018).

9. Karma Ura, Sabina Alkire, Tshoki Zangmo, Karma Wangdi, *A Short Guide to Gross National Happiness Index* (Thimphu: The Centre for Bhutan Studies, 2012), pp. 6–7. Available at: http://www.grossnationalhappiness.com/wp-content/uploads/2012/04/Short-GNH-Index-edited.pdf (accessed on 4 May 2018).

10. DP Kafley, 'Bhutan Ethnically Cleansing Native Hindu Population' (Vijayvaani.com, 20 July 2012). Available at: http://www.vijayvaani.com/ArticleDisplay.aspx?aid=2385 (accessed on 4 May 2018).

11. Cathy Scott-Clark and Adrian Levy, 'Bhutan: Fast forward into trouble' (*The Guardian*, 14 June 2003). Available at: http://www.theguardian.com/theguardian/2003/jun/14/weekend7.weekend2 (accessed on 4 May 2018).

12. Albert Gray and Harry CP Bell (trans and eds), *The Voyage of François Pyrard of Laval to the East Indies, the Maldives, the Moluccas, and Brazil*, two volumes (London: Haklutt Society, 1887). Available at: https://archive.org/details/voyagefranoispy01pyragoog (accessed on 4 May 2018).

13. Harry CP Bell, *The Maldive Islands: An Account of the Physical Features, Climate, History, Inhabitants, Productions, and Trade* (Chennai, New Delhi: Asian Educational Services, 2004).

14. John Kenneth Galbraith, *The Culture of Contentment* (London: Penguin, 1993), p. 173.

15. Ernest Braun, *Futile Progress: Technology's Empty Promise* (London: Earthscan, 1995), p. 37.

16. Richard Layard, *Happiness: Lessons from a New Science* (London: Penguin Books, 2005), p. 235.

17. Op cit. p. 275.

18. Ronald Wright, *A Short History of Progress* (Edinburgh: Cannongate, 2005), p. 13.

19. Gray, *Heresies*, pp. 2, 3.

20. Op cit.

21. Donella Meadows, Jorgen Randers and Dennis Meadows, *Limits to Growth: The 30-Year Update* (London: Earthscan, 2005), p. 1.

22. BCE: Before the Common Era.

23. Clive Ponting, *A New Green History of the World: The Environment and the Collapse of Great Civilisations* (London: Vintage Books, 2007), pp. 56, 57 and 69–72.

24. Wright, *A Short History of Progress*, p. 79.

25. CE: Common Era.

26. Jared Diamond, *Collapse: How Societies Choose to Fail or Succeed* (London: Penguin Books, 2011), pp. 176, 177.

27. Ponting, *A New Green History of the World*, p. 1.

28. For an alternative view of the fate of the Easter Islanders see Benny Peiser, 'From Genocide to Ecocide: The Rape of Rapa Nui', *Energy and Environment*, 16/3–4 (2005), pp. 513–539. Available at: http://journals.sagepub.com/doi/abs/10.1260/0958305054672385 (accessed on 4 May 2018).

29. Ponting, *A New Green History of the World*, p. 125.

30. 'For the first time in human history, food production will be limited on a global scale by the availability of land, water and energy'. Dr Fred Davies in 'Food shortages could be most critical world issue by mid-century' (*Science Daily*, 17 April 2014). Available at: http://www.sciencedaily.com/releases/2014/04/140417124704.htm (accessed on 4 May 2018).

31. 'Public Papers of the Presidents (January 20, 1949)' pp. 114–15. Quoted in Gilbert Rist, *The History of Development: From Western Origins to Global Faith* (London: Zed Books, 1997), p. 71.

32. The South Commission, *The Challenge to the South: The Report of the South Commission* (Oxford: Oxford University Press, 1990), p. 27. Available at: http://www.southcentre.int/wp-content/uploads/2013/02/The-Challenge-to-the-South_EN.pdf (accessed on 3 May 2018).

33. These definitions can sometimes be confusing as political and economic changes are always in flux. The term 'East–West' belongs to the *political* category in that it was used extensively to refer to the capitalist West, as representing Western Europe and North America, and the communist East, as representing Eastern Europe, Soviet Russia and China during the Cold War. The term 'North–South' belongs to the *economic* category and was given currency by the Brandt Report (Independent Commission on International Development Issues, *North–South: A Programme for Survival*). 'North' in this report referred to the rich countries located in the North, which included Western Europe, North America, Japan and confusingly Australia and New Zealand, and 'South' as including the poor countries in Africa, Asia and South America. Although there is considerable overlap in the use of these terms, in this chapter I shall be referring in the main to East–West in the political sense and North–South in the economic sense.

34. 'Final Communiqué of the Asian-African conference of Bandung (24 April 1955)'. Available at: http://franke.uchicago.edu/Final_Communique_Bandung_1955.pdf (accessed on 7 May 2018).

35. Independent Commission on International Development Issues, *North–South: A Programme for Survival* (London: Pan Books, 1980), also known as the Brandt Report.

36. The South Commission, *The Challenge to the South*, p. 10.

37. Op cit. p. 1.

38. Op cit. p. 10. For an extended analysis of the definitions of development, see Rist, *The History of Development*, pp. 9–24.

39. Wolfgang Sachs (ed.), *Global Ecology: A New Arena of Political Conflict* (London: Zed Books, 1993), p. 5.

40. Op cit. p. 5.

41. Denis Goulet, *Development Ethics: A Guide to Theory and Practice* (London: Zed Books, 1995), p. 41.

42. Amartya Sen, *Development as Freedom* (Oxford: Oxford University Press, 1999), pp. 3, 4.

43. Op cit. p. 15.

44. United Nations Development Programme, *Human Development Report 2016: Human Development for Everyone* (UNDP, 2016), pp. 198–201. Available at: http://hdr.undp.org/sites/default/files/2016_human_development_report.pdf (accessed on 5 July 2018).

45. World Commission on Environment and Development, *Our Common Future* (Oxford: Oxford University Press, 1987).

46. The article from which this quote is drawn was written in 1992 and world population passed the 7.6 billion mark in 2017.

47. Gretchen C Daily and Paul R Ehrlich, 'Population, Sustainability, and Earth's Carrying Capacity', *BioScience*, 42/10 (November 1992), pp. 761–771. Available at: https://morrisoninstitute.stanford.edu/publications/population-sustainability-and-earth-s-carrying-capacity (accessed on 7 May 2018).

48. Fred Spier, *Big History and the Future of Humanity* (Chichester: Wiley Blackwell, 2015), p. 303.

49. Op cit. p. 304.

50. 'Background' (United Nations: We can end poverty, undated). Available at: http://www.un.org/millenniumgoals/bkgd.shtml (accessed on 18 July 2018).

51. Shaohua Chen and Martin Ravallion, 'An update to the World Bank's estimates of consumption poverty in the developing world' (World Bank Briefing note 03-01-12). Available at: http://siteresources.worldbank.org/INTPOVCALNET/Resources/Global_Poverty_Update_2012_02-29-12.pdf (accessed on 7 May 2018).

52. UN General Assembly, 'Resolution adopted by the General Assembly on 25 September 2015' (United Nations, 21 October 2015). Available at: http://www.un.org/ga/search/view_doc.asp?symbol=A/RES/70/1&Lang=E (accessed on 7 May 2018).

53. United Nations, 'Sustainable Development Goals: Seventeen goals to transform our world' (UN, undated). Available at: http://www.un.org/sustainabledevelopment/sustainable-development-goals/ (accessed on 7 May 2018).

54. United Nations, 'Take Action! The lazy person's guide to saving the world' (UN, undated). Available at: http://www.un.org/sustainabledevelopment/takeaction (accessed on 7 May 2018).

55. Sidney Leng, 'China's dirty little secret: its growing wealth gap' (*South China Morning Post*, 7 July 2017. Available at: http://www.scmp.com/news/china/economy/article/2101775/chinas-rich-grabbing-bigger-slice-pie-ever (accessed on 15 July 2018).

56. Rukmini S, 'India's staggering wealth gap in five charts' (*The Hindu*, 8 December 2014, updated 23 September 2017). Available at: http://www.thehindu.com/data/indias-staggering-wealth-gap-in-five-charts/article10935670.ece (accessed on 7 May 2018).

57. Rupert Neate, 'Richest 1% own half the world's wealth, study finds' (*The Guardian*, 14 November 2017). Available at: https://www.theguardian.com/inequality/2017/nov/14/worlds-richest-wealth-credit-suisse (accessed on 7 May 2018).

58. The Islamic Foundation for Ecology and Environmental Sciences (IFEES). See www.ifees.org.uk.

59. Fazlun Khalid, 'Qur'an, Creation and Conservation' (IFEES, undated). Available at: http://www.ifees.org.uk/wp-content/uploads/2015/04/1380144345.pdf (accessed on 7 May 2018).

60. Wolfgang Sachs (ed.), *The Development Dictionary: A Guide to Knowledge as Power* (London: Zed Books, 1992), p. 1.

61. Henrietta Moore, Director of the Institute for Global Prosperity and Chair in Philosophy, Culture and Design at University College London. See, 'BBC

Radio 4: The End of Development'. Available at: http://www.henriettalmoore. com/2015/03/the-end-of-development-professor-henrietta-moore/ (accessed on 7 May 2018).

62. Interfaith Rainforest Initiative, 'Statement of Participants of the Interfaith Rainforest Initiative, 21 June 2017'. Issued by UNDP and the Norwegian Ministry of the Environment. For full statement see https://rfp.org/wp-content/ uploads/2017/09/Statement-of-Participants-of-the-Interfaith-Rainforest-Initiative.pdf (accessed on 8 June 2018).

63. Jonathan Watts, 'Brazil abolishes huge Amazon reserve in "biggest attack" in 50 years' (*The Guardian*, 24 August 2017). Available at: https://www.theguardian.com/ environment/2017/aug/24/brazil-abolishes-huge-amazon-reserve-in-biggest-attack-in-50-years (accessed on 1 May 2018).

64. WTO News, 'Azevedo: 2015 a pivotal year for trade and the environment' (WTO, 29 April 2015). Available at: https://www.wto.org/english/news_e/spra_e/spra56_e. htm. See also, audio link at: https://www.wto.org/audio/2015_4_28_dgspeech. mp3 (accessed on 7 May 2018).

65. WTO News, 'Azevêdo: WTO marks 20 years of helping boost trade growth (WTO, 1 January 2015). Available at: https://www.wto.org/english/news_e/news15_e/ dgra_01jan15_e.htm (accessed on 7 May 2018).

66. Szu Ping Chan, 'IMF: infrastructure spending spree last chance to revive growth' (*Telegraph*, 30 September 2014). Available at: http://www.telegraph.co.uk/finance/ economics/11128937/IMF-infrastructure-spending-spree-last-chance-to-revive-growth.html (accessed on 7 May 2018).

67. World Economic Studies Division, *World Economic Outlook, October 2017: Seeking Sustainable Growth* (IMF, October 2017). Available at: https://www.imf.org/en/ Publications/WEO/Issues/2017/09/19/world-economic-outlook-october-2017 (accessed May 2018).

68. WW Rostow, *The Process of Economic Growth*, second edition (Oxford: Oxford University Press, 1960), pp. 10, 108. (This observation first appeared in the 1952 edition not available to the writer.)

69. HV Hodson, *The Diseconomies of Growth* (New York: Ballantine Books, 1972), p. 31.

70. Price Waterhouse Cooper, 'UK Economic Outlook, 2018' (PWC, undated). Available at: https://www.pwc.co.uk/services/economics-policy/insights/uk-economic-outlook.html (accessed on 5 June 2018).

71. The line on the hull of a ship, also known as the International Load Line, indicating the legal limit of a ship's load.

72. Andrew Simms and Victoria Johnson, *Growth isn't possible: Why we need a new economic direction* (London: New Economics Foundation, 2010). Available at: http://b.3cdn.net/nefoundation/f19c45312a905d73c3_rbm6iecku.pdf (accessed on 7 May 2018).

73. Meadows et al., *Limits to Growth*, p. xv.

74. See, the Global Footprint Network.

75. Meadows et al., *Limits to Growth*, pp. 1–16.

76. Op cit. pp. 17–49.

77. 'World Footprint' (Global Footprint Network, undated). Available at: http://www. footprintnetwork.org/en/index.php/GFN/page/world_footprint/ (accessed on 7 May 2018).

78. Dan Strumpf, 'Dow Tumbles, Hitting 2015 Low on Growth Worries' (*The Wall Street Journal*, Europe edition, 21 August 2015), Vol. xxxiii, No. 141.
79. Tim Jackson, *Prosperity without Growth* (London: Earthscan, 2009), p. 65.
80. Op cit. p. 205.
81. Op cit. p. 189.

4

HOW HISTORY ACCELERATED

All the Time in the World

Transactions

I have been treading on the toes of historians in previous chapters and I will take a few more liberties in this one by looking at how the exchange of things between humans evolved. Succeeding civilizations would inherit legacies from previous ones until a seminal event occurred, providing history with a spurt that went on to influence other cultures; the invention of the wheel being a classic example of this. The invention of alphabets and numbers, which provided the foundations for human communication, also accelerated the transmission of cultural changes. Invasions like Alexander the Great's incursions into the East, which opened up trade routes, is another such example of a legacy that speeded up change. What all these different changes in history had in common was to give a push to another level of advancement within successively shorter timescales.

There was, however, one inevitable element of change that ran like a thread through all cultures, from the earliest times to this day, and that was the uniquely human habit of exchanging goods and services with each other. It began with the simplest of one-to-one transactions, and the emergence of money as an element of intermediation at a certain period in history accelerated this process. It may not look like it now, but the appearance of money way back in the mists of time was in itself a revolution. It was at the root of all the great civilizational advances we have made since then, but as is the habit of revolutions it has also had a disturbing effect on us all. As money evolved over millennia it also became part of the human genome, mutating into a technique that roused the human instinct for covetousness. Possessing objects of value increased a sense of self-worth. The dark side of the human being, in the shape of greed and the lust for power, fed the acquisitive instinct.

Barter implies a simple one-to-one exchange to the satisfaction of two individuals and that was the end of it. The intermediary quality of money changed the nature of this transaction, as the recipients of money in one transaction continue on to another transaction, and so on and on, using the same money in a continuous chain of transactions. In economic theory this is known as the velocity of money, and this is how one piece of paper with '$10' inscribed on it can accumulate many times its value in goods and services as it keeps changing hands. Considering that everything we do has a connection with the environment, we could conjure up an image whereby a pile of $10 bills would become a swarm of locusts devouring their way through fields of wheat. The British government has recently introduced long-life plasticized banknotes and it is likely that by the time one of these notes ends its life it will have been involved in tens if not hundreds of transactions. Thus one £10 note has the potential to purchase goods and services to the value of £1,000 and perhaps a lot more. However, money today has evolved into something more fantastic and of this more later (see 'Acceleration' below; also see chapter 1, 'A Lethal Cocktail').

We don't know if money was used in prehistoric times, but there were many forms of it in recorded history, depending on the culture they originated. One of the most famous of these pre-metallic currencies was the ubiquitous cowrie seashell. The 'cowrie was current over a far greater space and for a far greater length of time than any other'.[1] All past high civilizations developed metallic currencies and 'only the Incas of the Andes had managed to achieve a high degree of civilization without the use of money'.[2] Money was the life blood of civilizations that flowed through their arteries and it was to such an extent that it enabled societies to evolve within the carrying capacity of their local environment and spheres of influence.

I have attempted to emphasize the aspect of connectedness in this book, and the point is that every time money changes hands it sooner or later impacts on the environment. An interesting game would be to try and discover when it doesn't. We have managed to come through past millennia relatively unscathed, because of low population density and relatively modest resource demands, but this state of affairs was to change in the sixteenth century, which by the legitimization of usury/interest in Europe saw the climax of a movement that shifted the nature of transactions from the organic to the abstract.[3] This not only introduced a fundamental shift in the very nature of money, its use radically changed the way we relate to the natural world. It financed the impetus of modernity, so that we are now at the apex of a process defined as a consumer civilization. From a condition where we once functioned within the carrying capacity of a local environment, the human race has now collectively exceeded the carrying capacity of the entire planet. How did we get to this point? In this chapter I shall make an attempt to answer this question.

Gifts or barter?

We are social animals. Sharing, negotiating and transacting have been in our bloodstream from the times of our earliest ancestors. The family, the group and the clan had survival needs and their aspirations didn't go beyond food, clothing and shelter. Sharing roles and basic needs were crucial for survival, but this is not to say that the human fault lines of greed and selfishness didn't exist alongside the positive attributes of generosity and selflessness. Life was harsh, but this is how we describe it today as we have dealt with the challenges of the past – although we appear to have created some new intractable ones in the meantime. Harshness as we see it today was the norm in the past. For our ancestors, a bright outlook was immediate. It would perhaps have meant a good kill; some meat for the group garnished with some nuts they may have gathered, eaten with gusto around a fire in a cave shelter. Long term thinking did not exist. There were no pensions or insurance policies to be had. No savings banks. They lived for the moment and relationships were collaborative. There was a common pool of needs which were shared. Simple exchange of essentials came naturally – exchanging surplus meat for a stone axe, perhaps. These were simple transactions that occurred on the basis of need and mutuality. Anthropologists argue that a debt (not money debt as we know it today) culture was one of the means by which early humans conducted their exchanges, which are best described as 'owing favours'. These practices were encapsulated in the phrase, 'I am indebted to you for your favours', which we may have used on occasion ourselves. Or in the phrase, 'I am much obliged'. We could just as well refer to these exchanges as favour economies. Needless to say, intra and inter-group transactions had the potential for aggression, which was more often than not avoided by the use of subterfuge and bluff. What started off as a collaborative relationship for survival emerged as an exchange of favours, translated as gifts. Gifts then became an obligation that needed to be returned – it assumed the proportions of a debt. This is not the same as barter. This suggests a sort of negotiated interdependency at a time when the struggle for survival was at its harshest.

The conventional view encouraged by economists is that barter was the norm in the evolution of human transactions. But this does not appear to be the case according to the anthropologists. Barter became correspondingly more complicated in relation to the amount of things and labour services that were available in a given space–time situation. Negotiating exchange will have been difficult in early times, for how do you arrive at the relative value of wheat in exchange for salt for example? This is not to say that these kinds of exchanges did not take place, but barter came into its own only after what we understood by money came into existence, as money made it possible to estimate the relative value of the goods that were being exchanged; for example animal skins for grain. How else was it possible to determine the value of the number of skins in terms of a given weight or volume of grain? Anthropologists have been hard put to find barter economies in early societies.[4] They hold Adam Smith, the father of economics, responsible for perpetuating this myth in his seminal *Wealth of Nations* and, rather unfortunately, there does not appear to be any scholarly agreement either on the terminology applied to describe early human economic activity. One academic observes that, 'competitive gift exchange probably reached its most aggressive heights in the ritualised barter ceremonies among North American Indians'.[5] Which is it to be? Gifts or barter? Be that as it may, barter in its simplest form was not uncommon, but competitive gift exchanges suggests the existence of a power relationship. Here we need to recognize social status in these transactions, because the group or individual that ostensibly ranks high in this competition are the ones who are able to give away the most, leaving the receiver with a feeling of indebtedness. It could be argued that these rituals manifested power relationships based on generosity. This may explain why anthropologists were not able to find societies based on barter in the way Adam Smith and successive economists sought to define it.

Keeping accounts

It would have taken thousands of years to arrive at the kind of economic arrangements practiced by the Sumer civilization in Mesopotamia (see chapter 3, 'The Progress Trap') which has provided us with the earliest known records of any human civilization. By the fourth millennium BCE the Sumer civilization had a system of money based on silver. The bulk of the Mesopotamian cuneiform records archaeologists have unearthed were credit and debit entries maintained by priests and bureaucrats, and what we see here is the first step in the evolution of money as we know it today: 'The basic monetary unit was the silver shekel. One shekel's weight in silver was established as the equivalent of one gur, or bushel of barley. A shekel was subdivided into 60 minas, corresponding to one portion of barley – on the principle that there were 30 days in the month [another Sumerian innovation which has stayed with us to this day], and temple workers received two rations of barley every day.'[6] In this way a diligent Sumerian temple worker will have earned the equivalent of one silver shekel a month and this is one of the earliest records of the emergence of wage labour.

Trading records began to enter the human story in the third millennium BCE during the course of the Ebla civilization. It had its centre not far from present day Aleppo in Syria (tragically now a destroyed city) and it overlapped with the Sumerian civilization. Its prosperity was based on extensive trade, had a system of currency similar to that which the Sumerians had developed, and:

... operated on a gold and silver standard, and possessed huge quantities of both ... Commerce was priced in terms of gold and silver. Five bars of silver were pegged at a fixed and stable exchange rate to one bar of gold ... Although much trade still revolved around barter, the presence of an international gold and silver standard represents a major step towards a money economy.[7]

Eblaite merchants travelled far and wide and transported their merchandise in wagons driven by oxen. They created trading links with Afghanistan in the East and Egypt in the West. 'The taxes from one outlying district, sent to Ebla, include linen fabrics, top-quality fabrics, best quality multi-coloured dresses, sashes and tassels, hosiery'.[8] The Eblaites were prone to warfare and conquest and were unique in that they have been described by historians as the first recorded world power. And here we see, if not for the first time at least in earliest recorded history, the emergence of the combination of trade with aggression. War was (and is) good for business.

The basic elements that enabled inter-regime trade to take place almost 3,000 years ago was in place. These regimes were not entities we could refer to as nation states as we understand them today. They were defined by gods, culture, language, and ethnicity. They were power centres based on kings and queens; temples and priests; administrators and bureaucrats. Without this structure that enabled the keeping of records and the levying of taxes there would have been no way in which we would have known about these early societies in such detail. Borders were constantly changing and fluid. Markets that began as local celebrations that enabled the exchange of local produce evolved into inter-regime meeting places that dealt in exotic merchandise. This encouraged the mixing of people and made possible cultural exchanges. Dictionaries were being produced to enhance communication.

Gold at this point was largely held as a store of value and was uncommon as a means of exchange, for a gold-based coinage that could be used in everyday transactions had yet to be developed. Gold had a hypnotic effect on men and women from the earliest times in human history. So what is it about the metal that made it so valuable? The reality is that it has no intrinsic value. To prove this you need to go to a desert island with a bag full of gold bars. You can't eat it and all it will do for you is shine in the sunlight. You will find fish, fowl and fruit infinitely more valuable, because they will keep you alive. 'Stubborn resistance to oxidization, unusual density, and ready malleability – these simple natural attributes explain all there is to the romance of gold'.[9] It is rare, mined with difficulty, shines, lasts forever and can be shaped to anything that suits your purpose. What gives it value is that as a species we covet it and being in possession of quantities of it creates the illusion of power.

The pharaohs were casting gold bars as money in the fifth millennium, but the problems of purity, transportability and transactability were not resolved until about 700 BCE in Lydia (present day Western Turkey). The Lydians were reputedly the first people to mint gold and silver coins (the metal used for this purpose was in fact electrum, a compound of gold and silver).[10] A new breed of men we know today as traders and merchants were emerging. They were intermediaries; buyers and sellers whose main motivation was profit. We could say that this development had taken more or less 12,000 years to mature – during the course of the stone, bronze and iron ages – if we take the starting point of human settlements to be somewhere in the middle of the twelfth millennium BCE.

Trade evolves

Eblaite commerce was not unique in the same way as Sumerian agriculture was not the first to emerge in Neolithic times. As common sense would tell us, much the same was happening elsewhere with trade as with agriculture, and trade by its very nature involved others from distant places. But there were differences, if only ideological. Researchers have chosen to describe the Eblaite system as 'quasi-Entrepreneurial', the Egyptian as 'quasi-socialist' and the Mesopotamian as 'quasi-corporatist'.[11] We get a hint here in the entrepreneurial system that evolved in Ebla of merchants beginning to strike out independently, in contrast to their Egyptian and Mesopotamian contemporaries who were closely tied to palace and temple. We know about Sumer and Ebla in such detail because of their unique organization and the use of an evolved form of cuneiform symbols. This form of early writing enabled them to keep records, which we have discovered in archaeological digs. In the case of Ebla, 'a vast library of some 15,000 business documents was discovered on the site of Tell Mardikh in North West Syria, most in classical Mesopotamian cuneiform' as recently as 1975.[12] It is also not as if these developments happened suddenly, and at one and the same time certain cultures acquired the skills to maintain records. The processes that led to what we learn from records were gradual and organic, and traces of crude cuneiform writing have been found as far back as the eighth millennium BCE. The first phonemic alphabet itself did not appear until the Phoenicians developed it in the eleventh century BCE. The Greeks created the forerunner to the current Western alphabet in the eighth century BCE by adding symbols that represent vowels to the original Phoenician.[13] But the words you are reading now were printed on paper, itself a complex technology, which took a long time to develop. As we have seen, the Sumerians used clay tablets to maintain their accounting system; the Egyptians used papyrus in the fourth millennium BCE; the Chinese used bamboo in the second millennium BCE; early Buddhist texts were written on Ola palm leaves in the first century BCE; the original verses of the Qur'an were written on bone and leather; renewable wax tablets were also in use by different civilizations during different phases of their development. Paper as we know it today was invented in China during the course of the Han dynasty (206 BCE–220 CE). Muslims acquired this knowledge in the middle of the eighth century and by the ninth, Islamic Spain was printing Qur'ans and other books made from paper from where these techniques were passed on to Europe.

If Eblaite merchants were using wagons pulled by oxen loaded with merchandise travelling east and west then other traders must have been travelling to Ebla and elsewhere from these outlying regions. This is also an indication of the use to which the relatively new invention of the wheel was being put. The wheel made bulk transport a possibility, but the disadvantage of this mode of transport in those distant times was the absence of roads with hard surfaces that would have made the carriage of heavy loads feasible for long distances. Although it is known that paved roads existed in Ur in 4000 BCE, it is reasonable to assume that the Eblaite wagons did not reach mountainous Afghanistan and that much of these distances covered by the merchants were made possible by the use of pack animals such as horses, mules and the donkey. The camel was a late-comer. The Incense Road, from the Indian Ocean shores of Arabia that ran north to the Mediterranean, 'was made possible by the domestication of the camel in the late Second millennium BCE and the rise of caravan cities along the route such as Iathrib (Medina)'.[14] The two-humped camel 'came into general use for transporting goods'[15] about the same time in China. Given their capacity to traverse long distances and inhospitable spaces, both the Arabian camel and its double humped cousin further east were the 'ships of the desert' that opened up hitherto unknown land routes from one faraway place to another in those distant times. Camels then were responsible for linking the trade from the Incense Road, originating on the warm shores of the Indian Ocean, with the overland Silk Road, originating in distant China.

Civilizations began and evolved on the banks of great rivers and were the precursors of marine navigation. 'The first extensive trade routes ... up and down the great rivers which become the backbones of early civilizations – the Nile, the Tigris and Euphrates, the Indus and the Yellow River'.[16] The master navigators of the Mediterranean during the first and second millennium BCE were the Phoenicians, and they thrived in the region now defined by Lebanon and Syria. The Phoenicians built on the extensive maritime trade that originated between Egypt and Minoan Crete and carried it westward along the North African coast, and are known to have transported tin ingots from Cornwall in southern Britain. They also had control of the shipbuilding centre based in present day Cadiz in southern Spain.[17] We were to witness the sophistication of these shipbuilding techniques almost 700 years later in the voyages of the Iberian sailors that shrank the world to what it is today. The Phoenicians reached their peak in the tenth century BCE during the time of King Solomon in Israel.

Enter the Greeks

The first major intercontinental thrust of trade arrived with the exploits of Alexander the Great in the fourth century BCE. He had established himself within the space of a decade as the ruler of an intercontinental empire stretching from Egypt all the way to the Punjab in the Indian subcontinent, and 'By the end of the Hellenistic age, both the Indian subcontinent and China were slowly and tenuously integrating themselves into a hemispheric-wide trading system.'[18] Chinese silks were reaching Europe during this period in increasing quantities but it was not until the establishment of the Silk Road by the Han dynasty of China in the third century BCE that this trade really began to blossom. In this period, trade from the East was controlled by the Greek dynasties that succeeded Alexander: the Seleucids controlling the land routes with China and South Asia and the Ptolemaic dynasty 'controlling the western and northern end of the trade routes to Southern and South Asia [and] had begun to exploit trading opportunities in the region prior to the Roman involvement'.[19] The Greeks were overcome by the Romans in 30 BCE, but by this time they had left behind them a trade route that connected Alexandria in Egypt via the Red Sea to Aden and beyond.

Thus, as the Romans were beginning to build their vast empire, Europe was trading with distant China and all points in between by a land route defined by the Silk Road, and by a sea route that connected Alexandria in Egypt to Aden in Southern Arabia; and from there taking advantage of the maritime trade that had developed between the Arabs and the Indians. Arab sailors plied the Red Sea and Indian Ocean leg to ports on the Indian west coast and present day Sri Lanka, Indian vessels crossed the Bay of Bengal via the Palk Straits to Malacca at the southern corner of present day Malaysia, from where Chinese junks completed the link to mainland China.[20] During the Islamic period, Arabian seafarers were trading as far afield in what is now the Philippines by the eighth century CE.

This then was the pattern of trade that the Common Era inherited. It continued to expand against a backdrop of the growth of world religions, major wars, population disruption, constant boundary changes and internal adjustments as nations began to form, collapse and reform. Buddhism was growing in the East; Emperor Constantine of Rome converted to Christianity in 312 CE, from which point onwards began the gradual Christianization of Europe. The Holy Roman Empire held sway over Europe for 1,000 years, from 800 to 1806. By the seventh century, Islam had expanded from its original source in Makkah to the shores of the Atlantic in the West and the Chinese border in the East. The Crusaders conquered Jerusalem in 1099. The Mongols destroyed the Abbasid caliphate in Baghdad in 1258 and got as far as Poland on three occasions in the thirteenth century. Also by the end of the thirteenth century Kublai Khan, the grandson of Genghis Khan, was ruling China. The Ottoman Turks invaded Constantinople, ended Byzantine rule in 1453, and converted the magnificent Aya Sophia into a mosque. Seven hundred years of Muslim rule in Spain was ended by the Catholics in 1492, followed by the conversion of the great mosque in Cordoba into a church.

Account must also be taken at this point of the emergence of the magnetic compass. The magnetic properties of lodestone and iron were already known to the Chinese in the period before the Common Era and a compass-like device was used for divination purposes by the third century BCE Han Dynasty. By the twelfth century CE the Chinese had developed a compass sophisticated enough to be used for marine navigation. Given the trading relationships that had evolved between the Arabs and Chinese from the earliest times, it comes as no surprise that Arab sailors were using the compass for navigation purposes if not by the twelfth century at least by the early thirteenth, having taken this knowledge from the Chinese. We see the classic route of transmission of knowledge once again, when the Europeans learn of the use of the compass from the Muslim navigators. Another view suggests that the Europeans may have developed the compass independently, but the proximity of the appearance of the compass in Europe in the twelfth/thirteenth centuries would suggest that the upheavals in the Crusader period may have played a part in passing this knowledge on to Europe. Nothing surprising about this as conflict and the exchange of knowledge have gone hand-in-hand throughout history.

The Venetian connection

It had been a long haul from the earliest barter communities through the Mesopotamian civilizations to the apogee of Islamic dominance in the twelfth century, when by this time Muslim traders had established a form of proto-capitalism. And here we come across another example of the Islamic bridge (see chapter 1, 'The Bridge') in action between the East and the West. Muslim traders were an important conduit in east–west trade during the Middle Ages, and they maintained strong links with major Italian trading cities like Venice. There were two important outcomes for the evolution of European trade as a result of these connections, and one of these was the incorporation of the value transfer system known as the *hawālah* into Italian and French law. European traders also eventually adopted the practice of debt transfer practised by Muslim traders, which was forbidden by Roman law.[21]

> Venice became the 'liquid frontier' in trade between the Islamic East and the Christian West, often ignoring papal bans on trade with non-believers. So many fortunes in trade were made that a Venetian diplomat acknowledged the city's symbiotic relationship with Muslims. 'Being merchants we cannot live without them'.[22]

In this period trade expanded substantially, and it was largely a flow of goods from East to West purchased with the payment of gold and silver flowing in the opposite direction. Additionally, Venetians sold 'tin, and lead, as well as furs, linen and even hats in the East'.[23] Traders carried letters of credit and paper money and banks as we know them today did not exist. There was, however, one exception to this as far as it concerns paper money; it saw the light of day in China during the Tang Dynasty (618-907 CE), but these notes were issued by the state and bore no relation to modern paper currency (see 'Acceleration' below; also see chapter 1, 'A Lethal Cocktail').

Beginning with an estimated human population of four million in 10,000 BCE, numbers had risen by just a million to five million by 5,000 BCE. Approximately 5,000 years later, the population at the time of the birth of Jesus Christ had jumped to 170 million. In 700 CE, during the time of Prophet Muhammad, it was estimated at 210 million.[24] The environmental impact on the Earth by the end of the fifth millennium BCE would have been negligible, although agricultural activity was increasing in certain locations where water was available in copious quantities. Human population density was low and widely dispersed, but man's tendency to outgrow other species demonstrated the human capacity for aggression and adaptability.

Signs of Impatience

Enter the Portuguese

Trade between east and west, Islam and Christendom, grew inexorably. Aleppo, as the junction of the Silk Road from the East and the caravan routes from the South, continued to develop strong ties with Venice in the eastern Mediterranean. There were large colonies of Venetian merchants on the eastern Mediterranean coast and by the twelfth century Venice had become the most important trading link between Europe and the Islamic world. Aleppo continued to flourish during the time of the Ottoman caliphate, which had taken over the intermediating role between the East–West land trade; they were strategically placed to be able to do this. By the end of the fifteenth century, however, the Portuguese set about breaking the Venetian dominance of this trade and also wresting from the Arab navigators the monopoly they held over the sea routes. The responsibility, if not credit, for leading the way in accomplishing these tasks must go to the Portuguese navigator Vasco da Gama, who rounded the Cape of Good Hope in December 1497. In so doing he opened the sea route for European incursions to the East and its eventual domination.[25]

What followed was a textbook execution of what has now become known in the annals of European imperial history as gunboat diplomacy. Vasco da Gama impersonated a Muslim to gain an audience with the Sultan of Mozambique. The gifts he had brought were rejected by the Sultan, who was more sophisticated in his tastes than the Portuguese had imagined them to be. And when Vasco da Gama's ploy was discovered he had to flee to avoid the violence that was brewing. As he sailed out of the sultanate he treated the inhabitants of the city to a salvo of cannon fire. Sailing north alongside the East African coast he resorted to piracy and looted unarmed Arab trading vessels. He employed a local Arab pilot in Malindi (now part of Kenya) to take him to Calicut (Kozhikode in the present day state of Kerala) on the west coast of the Indian peninsula. He landed on the shores of the medieval Malabarian Kingdom ruled by the Samoothiri (anglicized Zamorin) dynasty in May 1498. Although he was initially welcomed by the Hindu king relations rapidly deteriorated and he bullied his way out of the kingdom, taking some hostages with him.[26] His pioneering sea journey was nevertheless a commercial success. Having established contact with India he returned home with enough cargo to pay his journey sixty times over. Within the space of five short years the Portuguese were in control of the Western coast of India, and to achieve this they had destroyed a centuries' old trading tradition between the Mediterranean, Arabs and Indians and much beyond. The bridgehead Vasco da Gama had established in the East was soon to be taken advantage of by the Dutch, the British, the French, the Spanish and eventually the Americans.

One of the persistent attributes on the negative side of the human equation is covetousness. It is usually the case that when there is a power equilibrium between societies, exchange and trade are the means by which one side acquires what it desires from the other. Although it grieved mediaeval Europeans to part with their silver, and sometimes gold, in exchange for Chinese silks they reluctantly continued with this trade. Unequal power relationships result in a mindset which produces undercurrents that manifest in greed, plunder and mayhem. And such was the case when Vasco da Gama sailed eastward in 1497, and Christopher Columbus westward, just five years previously. It was no longer necessary to part with silver or gold in exchange for anything. They could take what they wanted, including vast quantities of these very same precious metals, and return for more. The decided advantage the Europeans possessed above all else was an unassailable war technology in the form of the gunboat. A salvo of cannon fire, the planting of a flag, and vast open spaces, treasures by the shipload and 'savages' to be converted to Christianity were theirs for the taking. By the nineteenth century, the power nexus had swung so far in favour of the Europeans that the British were able to mount the infamous opium wars against the Chinese and extract land concessions from them. One of the prizes was Hong Kong. The French and the Americans played no small part in this escapade.

Chinese in the fray

For the Chinese, who invented gunpowder in the ninth century, to be made the victims of an invention they themselves pioneered 1,000 years earlier must have been a humiliating experience. Like paper and the compass, the transmission of this technology followed the same route: into Europe through the Islamic world. The Muslims acquired knowledge of gunpowder in the middle of the thirteenth century, and two centuries later the Turks had developed a super gun which was used to bring down the walls of Constantinople. The Europeans developed this technology to such an extent that by 'the seventeenth century their knowledge of gun founding and gunnery was one of the attractions which kept the Jesuit Fathers in favour with the Chinese authorities'.[27] But it proved of no avail as the Chinese, in spite of Jesuit know-how, could not prevent British gunboats from sailing up the Pearl River and taking Canton.

Unequal power relationships do not always have to end in violence and domination, as the Chinese had shown the world more than 600 years previously. The Chinese themselves were no mean sailors, and there is an episode in their naval history that is worth telling – if only as a salutary lesson for those wielding power. We can only speculate on how differently history might have turned out had the mindset of the Chinese rulers of the time been different and had their ambitions matched those of the Europeans in looking outwards. It would appear that their outlook has radically changed today, but this piece of naval history is an indication of the extent to which the sea routes between China, India, South Arabia and East Africa had developed in the centuries prior to the Portuguese incursions.

It was during the reign of the ebullient Ming dynasty ruler Zhu Di, who put together a vast fleet under the command of Admiral Zheng He (pronounced 'Her'), who was a Muslim of central Asian origin. He was captured during one of the innumerable clashes the Chinese had with the Mongols, duly castrated and employed by the emperor. He worked himself up the pecking order of the Imperial court, and described as a mountain of a man with a booming voice, was sent on diplomatic and trade missions by the emperor all the way to east Africa and points in between. His fleet consisted of more than 300 vessels, '30,000 sailors, seven grand eunuchs, and hundreds of other Ming officials, 180 physicians, five astrologers and ranks of geomancers, sailmakers, herbalists, blacksmiths, carpenters, tailors, cooks, accountants, merchants and interpreters'.[28]

> Viewed from the rocky outcrops of Dondra Head at the southernmost tip of Sri Lanka, the first sighting of the Ming fleet is a massive shadow on the horizon. As the shadow rises, it breaks into a cloud of tautly ribbed sail, aflame in the tropical sun. With relentless determination, the cloud draws ever closer, and in its fiery embrace an enormous city appears. A floating city, like nothing the world has ever seen before. No warning could have prepared officials, soldiers, or the thunder struck peasants who stand atop Dondra Head for the scene that unfolds below them. Stretched across miles of the Indian Ocean in terrifying majesty is the armada of Zheng He, admiral of the imperial Ming navy.[29]

It is reckoned that three of Vasco da Gama's ships could easily have been accommodated in the bowels of the Chinese admiral's vessel. Zheng He completed seven voyages, reminiscent of Sinbad the sailor who predated him in the annals of the Arabian Nights. He set sail on his first voyage in July 1405 and completed six more voyages, the last in 1431. In his final record he lists the thirty countries he had visited and, 'He writes of his efforts "to manifest the transforming power of virtue and to treat distant people with kindness"'.[30] It is unlikely that Vasco da Gama and those who followed him almost seventy years later had heard anything of this.

A new world in the making

We have now entered the phase of history which finds Europeans in a particularly effervescent mood; a cauldron of ideas, innovation, aspiration and enterprise is boiling over. They were well into the period usually referred to as the Renaissance and were going through the changes that would eventually bring them dominance over the rest of the world. Although the Chinese had known printing for centuries prior to the Renaissance, the prototype for the modern printing machine was invented by a German goldsmith in 1440. This quickened the pace by which knowledge was disseminated and a gradual but increasingly literate public was able to access material unavailable to previous generations. They were not only curious but were also characterized by an acquisitive, grasping streak that the world was about to experience. Also, while the Europeans coveted what the East had to offer there was no matching reciprocity by the East. The Hindu king of Calicut was as dismissive as the Sultan of Mozambique of the gifts Vasco da Gama had to offer. The European response was to impose themselves at their will and take what they wanted, for they had the means to do so with their superior weaponry.

There was also intense competition between the Europeans themselves for the riches they were able to seize from the rest of the world. The Portuguese had established a near monopoly over the European spice trade during the course of the sixteenth century, displacing the Venetians in the process. This also had an effect on Ottoman revenue as Aleppo, which was part of the caliphate, was beginning to lose its influence. Portuguese gunboats also made sure that Arab trading vessels no longer ventured into the open sea. The Portuguese monopoly was broken when the Dutch sailor Cornelis de Houtman sailed round the African Cape and reached Java in 1596.[31] British funded privateers (a euphemism for officially sanctioned pirates) arrived in the Malaysian Peninsula in 1598.

This was followed by an enterprise funded by the Dutch East India Company, which weighed anchor in Aceh in 1602.[32] This trading rivalry inevitably turned into colonialism and the fate of Ceylon (present day Sri Lanka) illustrates this point: the Portuguese began colonizing Sri Lanka in 1517; they were ousted by the Dutch in 1660, who ruled until the island nation became a British Crown Colony in 1803. There was no armed conflict between the Dutch and the British over the island, and the credit for this must go to the rivalries and conflicts generated in Europe itself. Napoleonic France was threatening the Dutch and the island figured as a pawn on a crowded chess board. It was treated like a piece of real estate (like Palestine and Jerusalem today) and it changed hands during negotiations which resulted in the Treaty of Amiens.[33] When the European conflicts were over the British, broadly speaking, had control over the Indian subcontinent and adjacent territories; the Dutch possessed the chain of islands that was eventually to become Indonesia; the Spaniards control of the islands that made up the Philippines, which was wrested from them later by the Americans; the French cobbled together the area defined as Indo-China. The

Portuguese were the losers in this, having just the crumbs to contend with in the south east Asian theatre for all the work they had done since Vasco da Gama rounded the Cape in 1497.

Much of the same was happening on the other side of the world. By the time the conquistadors had done their work and destroyed the Aztec, Inca and Mayan civilizations, amongst others, in South America, and the British, French, and Dutch explorers and the Pilgrim Fathers had done their work, the map of the 'New World' as we know it today began to take shape. The Spanish and the Portuguese divided America south of Mexico between themselves; the struggle between the British, French and Dutch for the north resulted ultimately in the birth of the United States of America and Canada, where the French gained a toe-hold in bilingual Quebec.

In Africa, the Portuguese and Spanish had established footholds on the west coast by the fifteenth century and by the nineteenth the Europeans had carved up virtually all of Africa between them. The only independent sub-Saharan African states left were Ethiopia and Liberia. Egypt and the territory north of the great Saharan barrier was a different matter altogether. They were part of the European experience from pre-classical times and when the colonial dust had finally settled down were shared between the British, French, Spanish and eventually the Italians.

From the sixteenth century onwards the Russians began their expansion eastwards towards the Pacific Ocean and by the end of the nineteenth century they had much of Central Asia under their control. As Tsarist Russia pushed towards the borders of Afghanistan, they hit the buffers of Imperial Britain. Russian ambitions to have the warm waters of the Indian Ocean lapping on the shores of their empire were thwarted by the British, who were protective of their Indian empire. Although the Russians – in their Soviet guise – were to enter Afghanistan in 1979, this was a short-lived enterprise as all those who sought to control Afghanistan, past and present, were to discover.

When we were discussing the Easter Islanders in 'The Progress Trap' in the previous chapter, we left them reaping the legacy of destruction they had sown for themselves after the Dutch sailors came upon them in 1722. Their subsequent fate provides a microscopic example of what was to happen to people who were 'discovered'. After the Europeans had done their exploring they were reduced to the state of a mouse being toyed with by a family of cats.

We have seen in the preceding section that Islamic trading practices such as money and debt transfer mechanisms were being absorbed by Venetian merchants in the twelfth century, which in turn enabled the growth of commercial or merchant capitalism (not to be confused with capitalism as we understand it today) in Europe.[34] Financing voyages provided sound – though occasionally risky – investments for European merchants, and monarchs played no small part in these enterprises. Thus, with big backers behind them the fifteenth century Iberian navigators initiated a process that within a short period of time destroyed a pattern of trade between the East and the West that had developed and matured over the centuries. The historical processes evolving in Europe at the time made these sea sagas inevitable. The pioneering navigators and those who were to follow earned themselves the admiration of their fellows, and have left an indelible place in history as remarkable seamen of their time. But it must still be said that the unmitigated violence that accompanied their seamanship was never justified, or even necessary, to achieve trading advantage. Good bargains can always be struck by a variety of other means – Zheng He had already shown the way. Yet the European sailors wanted more than a bargain, and the advantage they gained was by the use of paralysing fear. Why talk when you can just take? The people they 'discovered' had no idea how to cope with the threat, except to comply.

As we have seen, the world population grew from a mere 4 million in 10,000 BCE, rising to 5 million in 5000 BCE and 170 million at the time of the birth of Jesus Christ. By the end of the first millennium in the Common Era this had risen to 265 million and by 1500 the estimated global population stood at 425 million. This can be summarized as follows:

Estimated rate of growth (RG) of world population per millennium (pm)		
From 4m in 10,000 BCE	to 5m in 5,000 BCE	RG 200,000 pm
From 5m in 5,000 BCE	to 170m in 1 CE	RG 33,000,000 pm
From 170m in 1 CE	to 425m in 1500CE	RG 255,000,000 in half a millennium

From 1500 CE onwards the stage was set for a colossal transfer of global wealth and resources into Europe from the rest of the world. These events had no historical precedence and resulted in a world changed forever, establishing a western hegemonic power relationship over the rest of the globe that persists to this day.

Beginning with an estimated human population of just five million in 5000 BCE, it had risen to 425 million in 1500 CE. Population density was still low in spite of the disquieting growth. This growth can be attributed to the spread of agriculture, with a corresponding increase in activity involving the exchange of goods and services. Use of wind power was on the increase, as evidenced by successful sea voyages, and the use of domesticated animals combined with the wheel and the plough were the keys to this success. Although heavy demands were being made on finite planetary resources in certain isolated civilizations, overall environmental impact was manageable and within the carrying capacity of the planet.

A Spurt

Wealth flows one way

Silk and spices are freely available commodities in our high streets today. To put it like this is to state the all too obvious, but not so 500 years ago. These items were so coveted in fifteenth century Europe that status-wise they could be likened to owning an elite car in our time. What the encirclement of the globe brought to the Europeans were, amongst other things, gold and silver from South America; more gold, ivory and slaves from Africa; silver, myrrh, frankincense, and horses from the Middle East; yet more gold, pepper, cinnamon, and gemstones from India and Sri Lanka; cloves, nutmeg, and mace from the regions of Indonesia; silk, porcelain and lacquerware from China.[35] Although Spanish explorers failed to discover their mythical El Dorado in South America they did find gold by the boatload:

> A monumental mass of gold and silver sailed across the Atlantic from the New World to Spain during the 1500s ... the total European stock of gold and silver at the end of century was nearly five times its size in 1492. The volume was so enormous that the armed convoys that transported the treasure to Europe averaged about sixty ships; on occasion, the convoys included as many as one hundred ships. Each of these vessels carried over two hundred tons of cargo in the 1500s and around four hundred tons on larger ships in the 1600s. In 1564 alone, 154 ships arrived at Seville to debark their cargo of treasure. At the end of the sixteenth century, the precious metals accounted for the bulk of the value of everything shipped from America to Spain.[36]

The colonization of land included the cultivation of cash crops such as tobacco, sugar cane, tea, coffee and rubber which were exported in copious quantities to the 'mother' countries. One of the most infamous episodes of these colonial transactions was the Atlantic triangular trade, which lasted from the sixteenth to the nineteenth centuries. European ships sailed to Africa with beads, cloth and other manufactured goods. These goods were sold in exchange for slaves, who were taken on the second leg from the west coast of Africa across the Atlantic and sold to plantation owners in the Caribbean and the American colonies. These ships then returned to their home ports in Europe with cargoes of sugar, rum, and molasses from the West Indies and tobacco, hemp and cotton (a late-comer) from the American colonies. Handsome profits were made in this triangular trade, as it was then known, and a round trip took about three months to complete.

The reader will probably have heard of the story of the mutiny on the *Bounty*. The *Bounty* was a British naval vessel commissioned in 1789 to transport breadfruit plants from Tahiti in the Pacific Ocean to the Caribbean in the Atlantic. This was an example of the practice where flora and fauna were transplanted from one part of the colonized territories to another. Tea, coffee and rubber are the better known cash crops that were transferred for commercial gain in this manner. Tea originated in China and the British introduced the cultivation of this beverage in the Indian subcontinent to break this monopoly. It was a drink which all classes in Britain took to like ducks take to water and it also became popular amongst the colonized people themselves. Coffee originating in Ethiopia was subsequently cultivated in Yemen and drunk there in the Sufi lodges. It was passed on to the European world through Istanbul in the sixteenth century. By the middle of the seventeenth century coffee houses were thriving in London and the capitals of Europe, making this part of the commercial and intellectual landscape. It was not long before coffee was recognized as a lucrative source of profit. Coffee plants from Yemen were taken to Indonesia and subsequently to Sri Lanka by the Dutch; by the early eighteenth century Europeans were drinking Javanese coffee in their parlours. By the middle of the eighteenth century Brazil became one of the largest growers of coffee to flood the American market. We see signs of this in American western movies, where cowboys are frequently shown drinking mugs of coffee. Brazil was also the place where rubber was first discovered, where it grew wild in vast rainforests. On this occasion, the British replicated the enterprise shown by the Dutch in taking coffee plants to Java when a British explorer 'smuggled' rubber seeds out of Brazil and delivered them to the Royal Botanical Gardens in Kew, near London. Subsequently, these seeds were dispersed in the vast rubber plantations that were to open up in Malaysia, Sri Lanka and other tropical locations. This was a late nineteenth century escapade and coincided with the growth of the motor car industry, by supplying the rubber for tyres. These global enterprises, apart from bringing vast commercial advantages to the imperial powers, also radically and irrevocably changed the terrestrial and human ecology of the affected regions; not forgetting the wholesale destruction of tradition and cultures.

A glance at the map of the world in the early twentieth century would have shown North America in the north west, Australia and New Zealand in the far south east and South Africa and South America at the southern half, all populated by Europeans looking to start fresh lives away from their overpopulated countries; 'Before 1800, there was little European migration on a large scale except from the British Isles. Since then, something like 60 million Europeans have gone overseas'.[37] The region of Vladivostok on the eastern edge of Russia was ceded by China to their northern neighbour in the mid-nineteenth century and by the early twentieth was being populated by peasantry of Russian stock. Russian populations also moved south to the Central Asian regions. Large numbers of Chinese and Latin Americans settled on the west coast of the United States during the period of the California gold rush. Chinese also settled in British colonial territories like Malaysia and Singapore. They were welcomed for their work ethic and entrepreneurial spirit. For almost 100 years after slavery was abolished in 1833, over 3 million Indians were transported to various parts of the world as part of the indentured labour system. This was a form of debt bondage, and they were put to work in tea estates in Malaysia and Sri Lanka, sugar plantations in the Caribbean and employed as railway labourers in East Africa.

When the tasks concerning exploration, discovery and subjugation had overcome their initial impetus other matters relating to trade, exploitation and profit had to be addressed. Although force was not entirely ruled out as an option, the issue of trade needed to be prioritized over the long term with greater imagination. What grew out of this was a form of economic nationalism known as mercantilism, which thrived from the sixteenth to the eighteenth centuries. Mercantilism, which was an outgrowth of early commercial capitalism, aspired to bring large inflows of gold and silver into the national coffers and also maintain positive trade balances. One of the best known institutions that represented this model was the British East India Company. What was once the role of the military was now being passed on to big business, and here we see the earliest signs of the multinational corporation, now with armed mercenaries under their control.

Revolutions

In the meantime, other interesting things were happening at home that brought about a fundamental change in our relationships with the natural order. These seminal events were to take place in Britain: The introduction of crop rotation and better labour and land productivity from the middle of the seventeenth century created not only food surpluses but also surplus labour. This was the agricultural revolution, which was seen as a trigger for its industrial counterpart. A labour force was now available to contribute to the birth and growth of one of the epoch-changing events in history. The inventive genius of the British ushered in the Industrial Revolution that was set to launch a massive global upheaval. Eighteenth century inventions such as the spinning jenny, the water frame and the spinning mule revolutionized the manufacture of cotton textiles, introduced mass production and the factory system to the world. These events eventually destroyed the thriving Indian cotton industry.

James Watt invented the condenser that increased the power of the steam engine, making it possible to apply steam power for wider industrial use. The story goes that he discovered the power of steam while playing with a boiling kettle in his aunt's kitchen, but the fact is that the power of steam was known for over 1,000 years. There is, however, no reason to doubt this account as there are numerous coincidences throughout history of societies and individuals having chanced upon a particular innovation or invention independently of the other, although divided by time and space. For example, people in different periods of history developed ploughs to suit their particular soil and climate conditions; the power of steam was known since the time of the Greeks in the first century CE – there are accounts of an organ being powered by steam in twelfth-century France, of a rotating spit driven by steam in Ottoman Turkey in the mid-sixteenth century, and a patent being taken out by a Spanish engineer in the early seventeenth century for inventing a steam-powered water pump which drained mines of surplus water. Mechanical devices that incorporated wheels, pulleys, crankshafts and other devices had evolved out of man's ingenuity over the centuries, but the British inventors who used this past knowledge added some of their own original groundbreaking ideas to new devices and succeeded in creating an extraordinary period of invention previously unknown in history. Invention begets invention and it was not long before the steam driven locomotive emerged and changed the world forever. Watt's first working model of his version of the steam engine appeared in 1776, and a steam driven railway engine invented by Richard Trevithick was a working proposition by 1804. This revolution, however, was not just about new inventions that came thick and fast, the motivation behind this was financial reward. In a matter of less than fifty years there were over 2,000 miles of railway track carrying millions of passengers in Britain alone. We were also witnessing here the evolution of a new relationship between the scientist and the artisan which did not exist before, and this collaboration was soon to produce a new profession we know today as engineering, and a new specialism that had developed as technology with numerous sub-specialisms. There is a third and all important personality in this business, for it is a business, which is the financier; of this more shortly.

Within the space of three centuries Europe had turned the world upside down. It had grown in wealth, its population was spreading to all corners of the earth, and it had marshalled all the primary resources that the scouring of the Earth has brought to its feet. Where new needs became apparent, like rubber for example, it went about securing it from its primary source and growing and exploiting it in more accessible areas. It had the expertise and organizational capacity to do this and for example, it would set about destroying whole mountain ranges of rainforest if necessary to plant tea (see chapter 2, 'Rape of the Forests'). This form of global environmental destruction of flora and fauna was unprecedented. The power generated by this new wealth made possible the appropriation of wealth from other parts of the world.[38] Industrialization was thriving on its self-feeding capacity to open up and create new resources. A process I would describe as the double feedback loop developed, which ensured that wealth taken out of the colonized countries came back in part as manufactured goods, which were then sold at a profit and the proceeds returned to Europe once again. In the early decades of the last century, my grandmother consumed jam, butter, and biscuits sealed in air tight tins and my mother fed us with milk processed from milk powder and applied talcum powder on our bottoms, all from products labelled 'Made in Britain'. This is at the level of the mundane, but on a national scale Sri Lanka had a well-developed infrastructure: tramways in Colombo, its capital, streets lit by gaslight from gas supplied through a network of pipes processed from imported coal, all established and running by the early years of the twentieth century. Multiply these building initiatives to encompass the length and breadth of an empire upon which the 'sun never set' gives some idea of the scale and complexity of this colossal enterprise. Capital and labour costs were met from wealth generated from the previous exploitation of local resources and local labour, having the long term effect of sustaining industrial growth in the 'mother country', enriching it beyond its wildest dreams and keeping its labour force at work.

The damage caused to the environment by the Industrial Revolution was immense, and in most cases irreversible. It also affected traditional societies and industries. For example, the traditional Indian textile industry was a victim of the Industrial Revolution. Calico cloth was originally produced in Calicut, the same location which provided Vasco Da Gama respite from his first eventful sea voyage, but in the nineteenth century poor Indians were encouraged to buy cheap cloth mass-produced in England after British mill owners complained that imported cloth was taking jobs from British workers. Hence, they were now paying the wages of the mill workers in Lancashire, and contributing to the profits of the capitalist mill owners, instead of using products created by their own industries.

The human population grew from 425 million in 1500 to 720 million in 1750, at the beginning of the Industrial Revolution, an addition of 395 million souls in just 250 years.[39] This brief period matched the growth of population in the whole of human history prior to 1500 CE, just short of 35 million, a sure sign of things to come. The period from 1500 CE to the Industrial Revolution witnessed a massive transfer of wealth to Europe from the rest of the world. This was accompanied by reverse flows of population from Europe to both the east and the west. A parallel increase in resource extraction was threatening ecosystems, but environmental awareness was nowhere on the horizon. Coal was crowned king, though the Earth was coping reasonably well in spite of the huge wounds inflicted on its surface. The Earth's carrying capacity had yet to be breached and conservation of finite resources did not appear to be a matter of concern.

Acceleration

Fast forward

By this time the Silk Road, the Incense Road and the sea routes pioneered by the Arabs, Indians and Chinese had drifted into the mists of history. The industrial contagion was taking hold of Europe and spreading outward. Germany and France were closing in on the heels of Great Britain and industrialization rapidly spread onwards to the Americas and the new territories such as Australia and New Zealand. The Baltimore and Ohio Railroad, the first in the US, was built in 1828; the first railway in India was opened in 1853; in Australia in 1854;

in Japan, now part of the industrial race, in 1872; in China in 1876. One gets a sense that the planet was undergoing rapid and relentless change that was unprecedented and unstoppable. Having absorbed the revolutionary industrial impulses from Great Britain, the all-conquering European nations were now emulating the British experience in their own occupied territories. From now on these impulses, which were referred to as 'European' in the past, now have to be considered as 'western' because of the emergence of the United States of America as a major global player in the planet's history. It became the world's largest industrial country by the end of the nineteenth century.

How to describe events that were so violent and sudden? How to describe these events which were diabolically destructive of the earth in their impulses and were causing irreparable and irreversible changes to planet Earth – our only home – and at the same time creating the illusion of a golden age? There is a thread in the millennial events I describe in this brief narrative of a quickening of pace in human activity from one event to another. Writing may have begun as scribbles in the sand, evolving into cuneiform script, then sophisticated alphabets, sacred texts, manuscripts, books by the million, newsprint printed each day by the billion, culminating in the voice recognition software I am using to 'write' these words. Or consider the wheel, whose first known use was in Mesopotamia in the sixth millennium BCE for making pottery. Subsequently used in transport in the fourth millennium BCE by the Eblaites, and then rapidly adapted for use in war, milling grain and irrigation. So now it's wheels everywhere, to the point that our civilization would collapse without this invention. Steam was slow to be utilized as an energy source and after a gradual evolutionary process it was given a spurt in the eighteenth and nineteenth centuries. Or, to bring this idea up to date, consider the mobile phone. It emerged in the 1980s and now it seems, in the matter of thirty years or so, that we all use one – including my 10-year-old grandson. We are now in the middle of another revolution, that of technology, and we seem to be using the same language to describe this event as we used to describe its industrial counterpart 200 years ago. Klaus Schwab, the founder and executive director of the World Economic Forum, opines:

> We stand on the brink of a technological revolution that will fundamentally alter the way we live, work, and relate to one another. In its scale, scope, and complexity, the transformation will be unlike anything humankind has experienced before ... [A new] Industrial Revolution [that] is evolving at an exponential rather than a linear pace. Moreover, it is disrupting almost every industry in every country. And the breadth and depth of these changes herald the transformation of entire systems of production, management, and governance.[40]

We haven't yet learnt the lessons of the first Industrial Revolution.

I came across a unique description of these processes, the fourth Industrial Revolution notwithstanding, in my reading more than forty years ago. I have been writing about it ever since and in my view it cannot be repeated often enough because it gives us a clue as to what I consider to be the root cause of our present day dilemmas. It is particularly interesting that the proponent of these ideas was an American, writing at the time when America overtook the rest of the world as the foremost industrial power late in the nineteenth century. I am referring to the work of Henry Adams, the controversial scientist and historian who propounded a theory over 100 years ago which suggested that the acceleration of technological change was forcing the acceleration of history. Adams was both a scientist and a man of letters, and he attempted to formulate a theory of history that could predict the future: mathematize the historical process. We now call this futurology, and forecasting the future is a well-honed discipline; today, for example, we use computer modelling in forecasting the consequences of climate change. Adams' favourite tool was mathematics. He asserts that 'the acceleration [of society] in the seventeenth century, as compared with that of any previous age, was rapid, and that of the eighteenth was startling. The acceleration became even measurable, for it took the form of utilizing heat as force, through the steam-engine, and this addition of power was measurable in the coal output.'[41]

It will suffice to say that Adams constructed a graph to show that there was a relationship between the rate of consumption and use of energy, and what he described as 'technological progress'. The result was an exponential curve, the Law of Acceleration of history:[42]

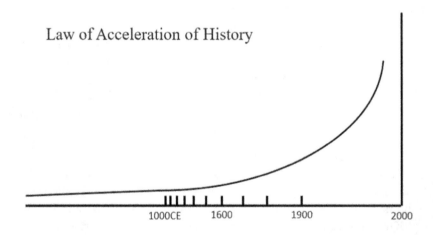

As Adams went on to say: 'The world did not double or treble its movement between 1800 and 1900, but measured by any standard known to science – by horse-power, calories, volts, mass in any shape, – the tension and vibration and volume and so-called progression of society were fully a thousand times greater in 1900 than in 1800'.[43] Furthermore, 'at the accelerated rate of progression since 1600, it will not need another century or half century to turn thought upside down. Law in that case would disappear ... and give place to force. Morality would become police. Explosives would reach cosmic violence. Disintegration would overcome integration.'[44]

Henry Adams, who lived most of his life in the nineteenth century, had no idea of the emergence of nuclear fission, as this was only discovered in 1938, and also said that we have until 2025 to turn things round. When he made these uncanny predictions, he had no knowledge of the emergence of the new nuclear physics and global warming as life-changing issues, but all of these forecasts have either already unfolded, or are in the process of so doing. The point of crucial interest for us, in this analysis, is the time at which the graph began to rise. Ever since the human family emerged from being hunter-gatherers to settled farming communities, human civilization advanced gradually over aeons of time. The acceleration graph shows an abrupt, upward trajectory from the sixteenth century onwards, and it is during this time that we can begin to discern the appearance of factors lying at the source of this sudden surge. Adams, in his historical cultural context, was only aware of what we now see as secondary and tertiary causes. What was important to him was technological change allied to energy consumption, and his valuable work 100 years ago produced this alarming exponential growth curve. A closer examination of this through an Islamic prism will, I believe, clearly show us the primary cause. What propelled the human race into an uncertain and unpredictable modernity was the legitimization of usury in Europe. I make no distinction between the terms usury and interest, and Margrit Kennedy demonstrates mathematically why this is the case, bearing in mind that all bank interest is compounded (also see chapter 3, 'The Fantasy of Growth'):

> Based on interest and compound interest, our money doubles at regular intervals, i.e., it follows an exponential growth pattern. This explains why we are in trouble with our monetary system today. Interest, in fact, acts like cancer in our social structure ... the time periods needed for our money to double at compound interest rates:
>
> - at 3%, 24 years;
> - at 6%, 12 years;
> - at 12%, 6 years.

Even at 1% compound interest, we have an exponential growth curve, with a doubling time of 72 years.[45]

Enter the banks

The banking system as we know it today began its evolution in northern Italy in the thirteenth century, but the English statute of 1545, enacted during the reign of Henry VIII, is 'of paramount importance in the history of usury and consequently in the history of money and banking also'.[46] The English monarch had the support of John Calvin, the puritan preacher and one of the leading thinkers of the Christian reform movement, who 'wrote a famous letter on usury and delivered sermons on the same subject'.[47] Usury was growing in Europe long before the sixteenth century Reformation. Fourteenth-century Florence was the financial capital of mediaeval Europe and even its secular authorities frowned at the practice of usury, in keeping with Christian teaching. In dealing with this matter it 'first prohibited credit transactions altogether, and then imported Jews to conduct a business forbidden to Christians'.[48]

Technically, in Judaism usury was only prohibited between Jews: 'Unto thy brother thou shall not lend upon usury, that the Lord thy God may bless thee in all that thou settest thine hands to' (*Deuteronomy* 23: 20). This enabled Jews to lend with interest to non-Jews, a practice which became one of the reasons for their unpopularity in the Middle Ages. Islam is more encompassing in its condemnation: '*Whatever you lend out in usury to gain value through other people's wealth will not increase in God's eyes*' (30: 39). Given that the evolution of the modern world occurred mostly under Christian influence, it is this religion's change of direction over time that is really most relevant to our purpose.[49]

The Christian prohibition on usury originates in Christ's action in the temple:

> 'Jesus entered the temple courts and drove out all who were buying and selling there. He overturned the tables of the money changers and the benches of those selling doves. "It is written," he said to them, "'My house will be called a house of prayer', but you are making it a den of robbers"' (*Matthew* 21: 12; New International version).

There was a great deal of soul searching in the Christian establishment before, during and after the time of Calvin on this matter. Calvin's position gave Henry VIII an opening to legalize usury in 1545, after his break with Rome. But it would appear that Rome itself has no position on this matter to this day, having left the biblical teachings on this matter to lapse: 'Attempts at finding an expert opinion among the Jesuits, Augustinians, Dominicans, Salvadorians, and even professors of moral theology in Third World seminaries teaching theology of economic justice failed to turn up anybody who remembered the forgotten Doctrine of Usury.'[50]

Much the same thing is happening in the Islamic world today. All Muslim countries, including those that describe themselves as Islamic, function within a system of financial intermediation not of their making. When the oil rich countries of the Middle East could have used their Islamic position to at least make an attempt to modify an iniquitous global system they opted to go along with the mainstream growth model. The Calvinists amongst Muslim scholars, supported by western-educated Muslim economists, have opened doors for Muslim rulers to enable them to join the global system with little perceived damage to their Islamic credentials, meaning that this is deceiving a lot of people. By and large, the rest of the scholarly fraternity have either tacitly accepted the status quo or ignored the issue altogether. Modernity is a tantalizing trap – the bait 'progress and development'. Leaving aside this politico-economic reality, soul searching does go on amongst Muslims about the issue of banks and usury. A minority of Muslim thinkers take a different view, and I am aware of a small group of UK-based intellectuals who have given this matter some considerable thought and whose workshop on the subject I was privileged to attend in 1987.[51]

The inequities to which people have been subjected by interest/usury have been recognized by past civilizations, and in the late second century CE (*c.*170), as the Han Dynasty was collapsing, a Daoist republic arose in what is today Sichuan Province in the West of China. Led by the Zhang family descendants of Zhang Dao Ling, who is often described as the founder of Daoism as a specific religion (*c.*140 CE), the Zhangs created an independent republic and ruled according to core Daoist philosophical beliefs. These were then translated into a handbook for how to govern called the *Taiping Jing* (the *Classic of Great Peace*). In this unique book, laws are set out in detail as to how to rule and this includes a complete ban on usury. The Zhangs' had to surrender around 220 to the rising power of a Daoist-inspired new kingdom called the Kingdom of Shu or Han-Shu, whose Prime Minister was a Daoist Master. The *Taiping Jing* survived and has informed informal governance by Daoists ever since.[52]

Muslim traders went to China before the Indian Ocean trade was usurped by the Europeans, and trade did thrive before the advent of the fractional reserve system. In this, the Islamic banks appear to have missed a point and are now wholly beholden to the global fractional reserve system, which as I write is now mutating into something even more fantastical. Banks today conduct their business in intangibles and we are now seeing the beginnings of a cashless society. Money, we are told, is going to be nothing more than a social relationship (see chapter 1, 'A Lethal Cocktail'). We are so dependent on banks today it is difficult for us imagine a time when there were none. There were such people as moneylenders in the past, hence the injunctions in the Bible and the Qur'an, and as we shall see presently, the coming of banks that rule all our lives is a recent historical phenomenon that radically altered the nature of money and our relationship to it. To be sure there are non-bank moneylenders around today, but they are a step or two removed from where the real mischief is taking place.

We discussed Rostow previously (see chapter 3, 'The Fantasy of Growth') and in his seminal work on the idea of economic growth. He observes: 'Virtually without exception, the take-off periods [of growth] have been marked by the extension of banking institutions which expanded the supply of working capital'.[53] This manifests an unquestioning acceptance of the banking model, which creates money out of nothing, has evolved since the Middle Ages in Europe and has now come to be deeply entrenched in our collective psyches. Even Karl Polanyi, in his acclaimed work on political economy, only makes glancing references to the subject. For example, 'With money, the threat was to productive enterprise, the existence of which was imperilled by any fall in the price level caused by use of commodity money'.[54] Nevertheless, economic textbooks discuss this subject and go to some length in describing the system, and what caught my attention was a couple of sentences in a book I purchased during my student days, which I still keep on my shelves: 'In fact, when a bank makes a loan, it does not usually lend actual cash. Instead, it gives the borrower the right to draw cheques upon it, although he has made no corresponding deposit.'[55] This observation provides us with a stepping stone for the discussion of money and banks. The bank had no money to give but the borrower nevertheless left the premises feeling that he had enough of it to meet his immediate needs.

Fantasy land

The banks have cast their spell on a grand scale. Even though they have been responsible for an unprecedented series of financial crises that climaxed in 2008 we appear to be still in their vice like grip. It is noted that greed, one of the human race's closest accomplices, has brought powerful economies almost to their knees and has affected people living in remote communities who haven't the slightest idea how their lives have been disrupted. Over-extended debt fuelling consumption patterns, and an economic growth model that is anything but sustainable, is now creating nightmare visions of global collapse. This is focusing people's attention on the banking system and the actual nature of so-called money as never before. It is now increasingly common to come across interesting appraisals of money like this, for example, from the prolific American economist and former Editor-in-Chief of the *Harvard Business Review*, Joel Kurtzman: 'in spite of all its fervid activity, money remains a naked symbol with no intrinsic value of its own and no direct linkage to anything specific'.[56] Money has come to be recognized as mere tokens and 'there is something quite magical about the way money is created. No other commodity works quite the same way. The money supply grows through use; it expands through debt. The more we lend, the more we have. The more debt there is, the more there is'.[57]

The money we have in our pockets is the direct result of the creation of a debt by someone, somewhere. It is not difficult to come to the conclusion from this description that the financial system, as we know it today, is a mirage and it is this mirage that rules our lives. The tokens of 'value' that Kurtzman describes are created by the banks legally from nothing, and grow exponentially *ad infinitum*. But we know that the natural world, which is subject to drastic resource depletion, does have limits and is finite. This equation is lopsided and poses this question: For how long can we continue to create this infinite amount of token finance to exploit the real and tangible resources of a finite world? Looked at from this perspective, money, as the modern world has contrived it, assumes the characteristics of a virus eating into the fabric of the planet. The consequences of this become increasingly visible in the form of climate change, massive global environmental degradation and unprecedented social disruption.

It is generally known that Islam prohibits usury or the taking of interest, and the term used in the Qur'an for this is *ribā*. This term has wide connotations and, simply put, it means one cannot bring something out of nothing. Thus, prohibition on *ribā* is also seen as prohibiting the creation of credit from debt. The Qur'an denounces these practices vehemently, and we can see why from the foregoing discussion: *Those who take ribā* (usury/interest) *will rise up on the Day of Resurrection like someone tormented by Satan's touch* (2: 275).

What Kurtzman has described so eloquently is known as 'fractional reserve banking', which originated in the principalities of thirteenth-century Italy. This system was entrenched by the creation of the Bank of England in an Act of Parliament in 1694. Uniquely, banks are the only institutions in the world that can charge its customers a fee (interest) for giving something they have created out of nothing and also demand collateral for it: 'The secret of creating money is being able to persuade people to accept one's IOU (a promise to pay in the future) as a medium of exchange. Whoever manages that trick can derive an income flow from the process (e.g., the medieval goldsmith's fees, or, today the interest on the loan that creates the money).'[58]

This illusion has now been entrenched by law in every single nation state and thus every self-respecting bank can perpetuate it by creating new money as it sees fit. How does this magic work? In Florentine Italy in the late Middle Ages, it was the habit of people with large quantities of surplus gold coins to deposit them with the local goldsmiths, who in return would give the depositor a receipt (IOU). When money was needed the depositor exchanged his receipts with the goldsmiths for what he required and the goldsmiths would charge a legitimate fee for this service. There were two unintended outcomes that resulted from this practice: As time went by depositors, instead of going to the goldsmiths to withdraw the coins, started to settle payments between themselves by exchanging these IOUs and paper money was born. The second outcome found the goldsmiths in an odd situation when they discovered that they had large deposits of dormant gold coins in their vaults over long periods of time, as not all the depositors wanted their gold coins out at the same time. The IOUs were functioning as a currency, with the sure knowledge in the minds of the people who were carrying out these transactions that they could be exchanged for gold coins at any time. This mindset is called confidence. Then one innovative goldsmith started what we know as banking today by issuing more IOUs than he had gold to people wanting a loan and charged interest for it. The risk the original goldsmith, and the evolving banking movement, ran was to issue more IOUs than it would be prudent to do because there wasn't enough gold in the vaults to honour all the receipts if all the depositors wanted their gold out at the same time. Over a period of time, however, it came to be recognized that if a certain fraction of gold was retained in the vaults this risk could be managed.[59] This was the origin of the fractional system. The first subterfuge was to create money out of nothing and the second was to charge a fee, interest/usury for lending that nothing.

The present system of financial intermediation that originated in Western Europe to run planet Earth could be likened to the cement that holds the house of cards together. But this cement is unstable. Unease and feelings of uncertainty prevail; there is the odd collapse here and a rumble there; breaches appear in the walls; politicians and bankers feverishly carry out running repairs; interest rates are manipulated; money is printed. What is to be prescribed between austerity and stimulus packages? Stock markets respond with alarming volatility; the G20 nations worry about the absence of growth; the IMF and the World Bank oblige by prescribing the growth medicine; China strives to keep its growth level above 6 per cent; the Japanese economy slumbers; India booms as does its poverty. The problem has been fixed – or has it? There is a breach in another wall in the house of cards and it is bigger than the last one.

The banking system supplies more props to support the structure. A sudden tug at these props will cause another global collapse and it is people living mainly in the 'advanced' world who will see drastic disruption in their lives. Some would argue that this wouldn't be a bad thing for global warming, as an event such as this would drastically reduce global carbon emissions. On the other hand, shoring the system up with yet more props will increase the problems we are now witnessing; more stimulus packages; yet more money being printed; yet more global warming. Whither sustainable development? There is another crash to come, either way – and a big one at that.

There was once a natural process in history: Thought stacked on thought stimulated further thought down the ages; innovation stacked on innovation stimulated further innovation; invention stacked on invention stimulated further invention; cultures stacked on cultures stimulated more culture; civilization stacked on civilizations stimulated the further growth of civilization. As but one example of these processes, the historical record shows how Muslims acted as a bridge between China and Europe in the manner in which paper, the compass and gunpowder found their way west. There has been a progressive contraction of time intervals between each of these stages; originally measured in terms of millennia, then in terms of centuries, then decades and now things change almost overnight. Somehow, common sense does not appear to stack up on common sense, wisdom on wisdom and experience on experience. The most profound change occurred in pre-modern Europe when in the soul searching over usury there was a shift in basic human transactions from fact into fiction; from the tangible to the intangible; from the credible to the incredible.

The accelerant of history was usury. We have looked at how money evolved and for millennia the nature of human transactions was based on money, which I have described as organic, which in Islamic parlance could be described as *fitrah* money (see next chapter). Muslim traders based their transactions on this system at the height of Islamic civilization and wielded considerable influence on the evolving European trading system. Civilizations waxed and waned, populations grew and trade blossomed. In pre-modern times the process was gradual, as demonstrated by the Acceleration Curve. The fundamental shift that ensued in the cauldron of change in Europe was the legitimization of usury in the English legal system in the sixteenth century. It was an act of expediency but its profundity was not realized at the time. The rest, as it is often said, is history.

Adam Smith, the father of economics, went to great lengths to define money, as he understood it in the eighteenth century, as a commodity. Now economists are coming round to thinking that the money we use today is merely an information exchange system. Following this logic, money could legitimately be called 'energy'. Or perhaps it is the conduit for transforming this energy into our modern lifestyles that is having such a devastating effect on the environment. Or consumerism? It has also been described as a cancer, which usually has the habit of growing until it kills its host. I see money as a virus bred by the banks in their hot houses, which if allowed to multiply will gnaw its way through the fabric of the natural world until there is nothing left (see chapter 1, 'A Lethal Cocktail'). The money we use today is at the root of our environmental crisis.

Endnotes

1. Davies, *A History of Money*, p. 35. The cowrie is a seashell and the largest deposits of it were found around the Maldive Islands.
2. Op cit. p. 47.
3. I take the view that the terms usury and interest are interchangeable. To say otherwise is like trying to convince a teetotaller that having a glass of wine is more acceptable than consuming a whole bottle of it.
4. David Graeber, *Debt: The First 5,000 Years* (New York: Melville House Publishing, 2011), pp. 21 and 43.
5. Davies, *A History of Money*, note 1, p. 11.
6. Graeber, *Debt*, note 4, p. 39.
7. Karl Moore and David Lewis, *The Origins of Globalization* (New York: Routledge, 2009), pp. 35, 36.
8. 'The archive of Ebla: 2500–2250 BC' (History World, undated). Available at: http://www.historyworld.net/wrldhis/PlainTextHistories.asp?historyid=106 (accessed on 7 May 2018).
9. Bernstein, *The Power of Gold*, p. 4.
10. Op cit. pp. 24–32.
11. Moore and Lewis, *The Origins of Globalization*, p. 41.
12. Op cit. p. 34.
13. The evolution of writing is a complex subject. I have merely tried to record here the time it may have taken the human race from making scratchings on clay tablets – understood only by the writer and a handful of others – to the appearance of an alphabet understood by a wider group of people. The origins of writing in Egypt, Mesopotamia and China in their various and unique forms can be traced as far back as the fourth millennium BCE.
14. Richard L Smith, *Premodern Trade in World History* (London: Routledge, 2009), p. 101.

15. Op cit. p. 122.
16. 'History of Trade' (History of the World, undated). Available at: http://www. historyworld.net/wrldhis/PlainTextHistories.asp?historyid=ab72 (accessed on 7 May 2018).
17. Moore and Lewis, *The Origins of Globalization*, note 11, p. 109.
18. Op cit. p. 175
19. David Stone Potter, *The Roman Empire at Bay: AD 180–395* (London: Routledge, 2004), p. 20.
20. 'The Ancient Maritime Sea Route: 250 BC–250 AD' (Nabataean History, undated). Available at: http://nabataea.net/msroute.html (accessed on 7 May 2018).
21. Gamal Moursi Badr, 'Islamic Law: Its Relation to Other Legal Systems', *The American Journal of Comparative Law*, 26/2 (1 April 1978), pp. 187-198.
22. Anthony N Penna, *The Human Footprint: A Global Environmental History* (Chichester: Wiley Blackwell, 2015), p. 237.
23. Op cit.
24. 'World Population' (World History Site, undated). Available at: http:// worldhistorysite.com/population.html (accessed on 7 May 2018).
25. Diffie and Winius, *Foundations of the Portuguese Empire, 1415–1850*, vol. 1, p. 177.
26. MGS Narayanan, *Calicut: The City of Truth* (Calicut: University of Calicut, 2006).
27. Roberts, *The Pelican History of The World*, p. 603.
28. Frank Viviano, 'China's Great Armada', *National Geographic*, 208/1 (July 2005), p. 35. Available at: https://business.highbeam.com/5308/article-1G1-136122559/china-great-armada-six-hundred-years-ago-china-admiral (accessed on 7 May 2018).
29. Op cit. p. 34.
30. Op cit. p. 52.
31. Ames, *The Globe Encompassed*, p. 99.
32. Op cit. p. 144.
33. The Treaty of Amiens was concluded between the French Republic and Great Britain. It was signed in the city of Amiens on 25 March 1802.
34. Jairus Banaji, 'Islam, the Mediterranean and the Rise of Capitalism', *Journal of Historical Materialism*, 15/1 (2007), pp. 47-74. Available at: http://eprints.soas. ac.uk/15983/1/Islam%20and%20capitalism.pdf (accessed on 8 May 2018).
35. Ames, *The Globe Encompassed*, p. 2.
36. Bernstein, *The Power of Gold*, p. 135.
37. Roberts, *The Pelican History of the World*, note 27, p. 747.
38. Op cit. p. 739.
39. 'World Population' (World History Site, undated). Available at http:// worldhistorysite.com/population.html (accessed on 6 June 2018).
40. Klaus Schwab, 'The Fourth Industrial Revolution: what it means, how to respond' (World Economic Forum, 14 January 2016). Available at: https://www.weforum. org/agenda/2016/01/the-fourth-industrial-revolution-what-it-means-and-how-to-respond/ (accessed on 8 May 2018).
41. Henry Adams, *The Degradation of the Democratic Dogma* (New York: Peter Smith, 1949), pp. 290-291. This essay, 'The Rule of Phase applied to History', was first written in 1909.

42. This graph is modelled on the one that appears in Gerard Piel, 'The Acceleration of History', in Ritchie Calder and Hossein Amirsadeghi (eds), *The Future of a Troubled World* (London: Heinemann, 1983), p. 11.

43. Adams, *The Degradation of the Democratic Dogma*, p. 303

44. Adams, in a letter to Henry Osborn Taylor (1905), quoted in Piel, p. 13.

45. Margrit Kennedy, *Interest and Inflation Free Money* (Seva International, 1995), pp. 6, 7. Available at: http://userpage.fu-berlin.de/~roehrigw/kennedy/english/Interest-and-inflation-free-money (accessed on 28 May 2018).

46. Davies, *A History of Money*, p. 221.

47. RH Tawney, *Religion and the Rise of Capitalism: A Historical Study* (London: Penguin Books, 1990), p. 91.

48. Op cit. p. 49.

49. Bernard Lietaer, *The Future of Money: Creating New Wealth, Work and a Wiser World* (London: Century, 2001), p. 48.

50. Op cit. p. 48.

51. Orr and Clarke, *Banking*.

52. I am indebted to Martin Palmer the General Secretary of the Alliance of Religions and Conservation for this information.

53. Rostow, *The Process of Economic Growth*, pp. 10, 108.

54. Karl Polanyi, *The Great Transformation: The Political and Economic Origins of Our Time* (Boston, MA: Beacon Press, 2001), p. 204.

55. Fredric Benham, *Economics: A General Introduction* (London: Pitman, 1960), p. 432.

56. Joel Kurtzman, *The Death of Money: How the Electronic Economy Has Destabilised the World's Markets and Created Financial Chaos* (Toronto: Little Brown & Co., 1993), p. 71.

57. Op cit. p. 85.

58. Bernard Lietaer, *The Future of Money - Creating new wealth, work and a wiser world.* (Century: London 2001) p. 304.

59. Richard J Douthwaite, *The Ecology of Money* (Totnes: Green Books, 1999), p. 17.

5

ISLAM AND THE NATURAL WORLD

Mobilizing Faith

Biodegradable Earth

While human beings have not always been kind to nature, belief in the presence of the Divine and a sense of the sacred has had the effect of defining boundaries which were deemed uncrossable. People of previous times, that is before the belief took hold that the natural world was there to be plundered, were basically no different from ourselves. They had the same positive and negative human attributes, but the impact of human profligacy was contained within the natural order of things, which transcended technological and political sophistication. Excess in the natural order was contained because it was biodegradable. Until quite recently the human race, whether as rebels or conformists, ignorant or enlightened, in small self-governing communities or vast empires, as barbarians or the highly civilized, functioned naturally within unwritten boundaries. This existential reality was neither idyllic nor utopian. However expressed, all the major or minor religions and traditional cultures, for example, the Native American or Aboriginal Australian, were deeply rooted in the natural world and drew all their inspiration from it. When old civilizations, however opulent, profligate, greedy, or brutal died, the forest just grew over them or the sands covered their traces. They left no pollutants, damaging poisons or nuclear waste.[1]

In cosmological terms, human beings have only been around on planet Earth for the blink of an eye, and I have attempted to show in previous chapters how we have managed in about 500 years to unsettle the Earth's natural systems. We have done this to the extent that we have put ourselves at risk and also induced the mass extinction of other species. 'Commercial capitalism',[2] which had its beginnings in the twelfth century, had taken hold of Europe by the middle of the fifteenth. This had by then evolved sufficiently to finance the Iberians to sail west and east, followed soon after in the sixteenth century by the British, Dutch and French. The foundations of the world order we experience today were laid down by these events, which underpinned a culture of plunder. The ravaging of the Earth on a grand scale had begun, compounded by the emergence of industrial capitalism by the middle of the eighteenth century. 'We are [now] facing a critical moment in Earth's history as our overextended human presence is affecting every region of land and water.'[3] Exponential population growth alongside alarming resource depletion accompanied by heightened global pollution is the end result of a socio-economic model that thrives on intensified consumerism. It is also severely damaging the legacy we are leaving for future generations.

It was not only the natural world that was coming under attack. The social milieu in which faith and traditional communities had taken centuries to evolve were also simultaneously destroyed by these events. As the Europeans sailed out of their ports they carried with them, in addition to their cannon, a civilization that was itself in the throes of change, and as it changed the world changed with it. The motivation was supplied by the likes of Prince Henry the Navigator (see chapter 1, 'Hegemony'), Thomas Babington Macaulay (see chapter 1, 'Brains, Black Holes and the Enlightenment'), the Papal Bull (see chapter 1, 'Hegemony') and numerous others. This in the end caused greater devastation than all the gunboats put together. The assault was spearheaded by the introduction of a financial and economic model underpinned by an education system initially run by the mission schools that produced compliant colonial servants and markets for industrial goods. Now the western system of education has well and truly been embraced by the so-called developing world, the citizens of which are blithely unaware of the fact that this path to global integration is now leading to planetary disintegration. Ways of life that had evolved over the centuries were destroyed and belief systems that carried strong environmental teachings reduced to empty ritual. Having undergone a transformation in the colonial process described as secularization, faith communities who now embrace a global materialist culture are expected to contribute to the business of damage limitation.

How can people of faith, in a milieu that is essentially alien to them, contribute to the mitigation of these trends if not reverse them? We have lessons to learn from the past as we face planetary threats that do not distinguish between religion, race, nationality, rich, poor, the powerful, the weak, old and young. By bringing their spiritual and moral authority to bear, faith communities have much that is positive to contribute, and their awakening to the environmental crisis has had a long period of gestation going back over fifty years. The philosopher Seyyed Hossein Nasr[4] was writing on this subject from an Islamic perspective as early as 1968, and literature produced in this genre from all the traditions have been on the increase, with the eloquent perspectives of Christian eco-theologian Thomas Berry[5] among them. Then there was White's *The Historical Roots of Our Ecologic Crisis*, published as far back as 1967 (see chapter 1, Brains, Black Holes and the Enlightenment). It would seem that protective voices expressing concerns over man's relationship with the natural world are now being increasingly heard, though it must be said that it is not nearly enough.

A vital alliance

Multi-faith environmental activism was given a boost by Prince Philip in the 1980s, while he was president of the World Wildlife Fund (WWF, later renamed the World Wide Fund for Nature). He brought together world religious leaders in 1986 in Assisi, Italy, out of which emerged the Assisi Declarations on Nature. There were contributions from leaders of the five major religions of the world – Buddhism, Christianity, Hinduism, Islam and Judaism – laying down their respective approaches to the care of nature. The Muslim Declaration on Nature was delivered by Abdullah Omar Naseef,[6] who was at that time Secretary-General of the Muslim World League (Rabitat al-'Alam al-Islami). Six other faith traditions have since joined this initiative.

During this period I was invited by Martin Palmer, who ran the International Consultancy on Religion, Education and Culture (ICOREC) and was pioneering faith-based environmentalism, to join his project as Islam consultant. Palmer's organization was the engine behind Prince Philip's faith-based environmental activism. One of the first fruits of this collaboration was an invitation to co-edit *Islam and Ecology*[7] which was part of a major series on faith and ecology sponsored by the WWF.

I was subsequently invited to chair an international workshop of faith-based environmental activists and organizations in Japan in April 1995, which had as its objective the discovery of common ground between nine different faith traditions from east to west. The outcome of this was the Ohito Declaration for Religion Land and Conservation (ODLRC),[8] named after the town where this gathering took place. This workshop produced ten common environmental principles and ten areas of action where the faiths found common ground. This was part of a wider initiative held under the auspices of WWF, ICOREC and the Mokichi Okada Association (MOA).[9] An extension of this gathering, reconvened in England a month later, established the Alliance of Religions and Conservation (ARC) and empowered it to act as the hub of faith-based environmental activism and promote ODRLC. Palmer, who had coordinated these initiatives, was designated ARC's General Secretary. Having chaired the workshop that produced this declaration, I was invited to promote it internationally. I thus slipped into the role of a roving ambassador for ARC and conducted this work for over five years.

Promoting ODRLC involved a period of intense and tiring travel, but what made this effort worthwhile was the wide range of people and perspectives I encountered. Much fruit can be harvested by listening to people from other places and persuasions and it had nothing to do with the old cliché that all religions have much in common, although this does not seem to have made much difference over the centuries: 'The two truths I discovered during this time were that the human race is faced with a common threat of unprecedented proportions and that we, the human race, are the threat itself.'[10] There was also an emerging consensus about how profoundly unrealistic modernity is and that some if not all the solutions would come from an understanding of the disconnection between our deeply held inner beliefs and our outer behaviour. It was also a salutary experience to discover that here was an issue which, although self-induced, can bring people together unlike any other.

Leaders of the world's religions met again to pledge 'Sacred Gifts to the Earth', in Kathmandu, Nepal in 2000. This was a joint initiative by WWF and ARC to encourage, secure and celebrate significant new pledges by the world's major religions for the benefit of the environment and the natural world.[11] Muslims were represented by the Islamic Foundation for Ecology and Environmental Sciences (IFEES/*EcoIslam*) and two gifts from the Muslim world were presented at this gathering. The first of these was the Misali Island Marine Conservation project based in Pemba, Zanzibar, and run by the Misali Island Marine Conservation Association.[12] Ali Khamis Thani, who was appointed to manage the project, was possibly the first person to work in the capacity of an Islamic Environmental Officer in this emerging specialism. The second was the Jab'al Aja' Biodiversity conservation project located in the Hail region of Saudi Arabia. This project was nursed by Othman Abdur Rahman Llewellyn of the Saudi Wildlife Commission (SWC) based in Riyadh, Saudi Arabia,[13] who can claim credit for pioneering the reintroduction of traditional Islamic environmental practices.

In academia, I was invited to act as Islam consultant by Mary Evelyn Tucker and John Grim,[14] who were organizing a series of conferences on religion and ecology at Harvard University in the late 1990s. This was followed by a series of books based on these gatherings and *Islam and Ecology: A Bestowed Trust*[15] emerged from this in 2003.

A watershed event took place in New York in the weekend prior to the United Nations Climate Summit, which was held on 22 September 2014. More than 200 religious and spiritual leaders from across the world gathered there for the Religions for the Earth conference. This event was assembled by Karenna Gore at the Union Theological Seminary, New York and in the words of the organizers: 'This is no ordinary conference: as the world's political leaders prepare to address an unprecedented moral crisis, Union will offer a unique platform for the world's ethical leaders to voice the concerns and commitments of the spiritual and faith traditions.'[16]

Religions for the Earth served as a launching point for a multi-faith, global movement and GreenFaith[17] emerged out of this. Coordinated by Fletcher Harper, it aimed at raising the voices of diverse faith communities internationally in support of strong climate action. The following final lines of the Declaration for an Awakened Kinship with Earth conveys a sense of the profound seriousness by which Religions for the Earth is addressing this matter:

> '... we pledge ourselves as leaders of religious, spiritual and environmental communities to care for Earth as the stunning gift of God, a sacred trust.
>
> We pledge every effort to love our Earth home.'[18]

Pushed by climate change

The approach to the climate change conference of parties (COP 21) in 2015, which eventually produced the make or break Paris Agreement, generated a great deal of activity amongst the faith communities. The field was led by Pope Francis and the Vatican, who produced the encyclical known as *Laudato si'*[19] published in May 2015, which addressed the malaise of our times. It was a universal appeal that called for urgent action and for human sanity to prevail. It took three years to produce and the drafting committee consisted of three Noble laureates. This was followed by the Islamic Declaration on Global Climate Change[20] (see appendix) which was launched in August 2015, followed by the other major traditions expressing their concerns about the state of the planet.

More than thirty years have elapsed since Assisi and twenty since Japan, and much has happened in between. Although supported by secular agencies like ARC, WWF and IUCN (International Union for Conservation of Nature; Union Mondiale pour la Nature) faith communities are still at sea over this matter. They still lack coherence, but it is a growing movement of global proportions like no other and their voices are rising and demanding to be heard. With their vocal engagement in the climate change movement they are now becoming known as FBOs (Faith Based Organizations) – courtesy of secular bureaucrats – alongside NGOs, their long serving counterparts. There is an emerging consensus that there has been a profound shift in the societal structure within which faith groups have traditionally been used to functioning, and this allows us to engage in this area of concern with greater unanimity. There is also an acceptance that having lost the power they once wielded they are now just one player amongst many with fierce competing interests, but their strength lies in the networks they have been building over the centuries. The Qur'an observes:

> *Had Allah willed He would have made you a single community.*
>
> *But He wanted to test you regarding what has come to you.*
> *So compete with each other in doing good (5: 48).*

The nature of this competition is such that there can be no losers, and much that is positive emerges from these movements, not least of which is building understanding between people based on commonality of purpose, hopes and anxieties. This in my view is the best thing that has happened in today's climate of change scenario, but the fundamental processes that have brought our current civilization to its near terminal state seem to evade our notice. There continues to be an assumption made by nearly everyone who engages at this level that voicing concern and producing an ethical basis for action is the way forward. This is true only up to a point. While we need to spread the 'good word', having lost a sense of the Divine and desacralized the natural world to the extent that we have lost all connection with it, we have much to do to leave a habitable world for the generations that follow us. Whether the major faiths establish new eco-narratives or rediscover the old ones, or we bring to bear the wisdom of the indigenous traditions, we are all without exception seduced by the trap of consumerism, at the base of which is a political economy to which we all subscribe. This has been discussed in previous chapters and if this is not addressed urgently and seriously then no new narrative or revival of old or traditional wisdom can save the planet. We need to go beyond tinkering with technicalities which in spite of its laudable objectives the COP process is a part. Time now to get to the root of the matter.

The Muslim contribution

When I embarked on this journey in the 1980s, I discovered a group of people looking at these issues from a broadly based faith perspective with whom I was able to relate. Given that we were collectively responsible for leaving a devastating legacy for future generations regardless of any religious, national or cultural affiliation, it made sense to work with people of other persuasions to return Earth to some kind of equilibrium. My activities brought me in touch with the likes of the Schumacher Society, Friends of the Earth and the International Consultancy for Religion Education and Culture (subsequently the Alliance of Religions and Conservations), with whom I functioned as Islam consultant for over ten years.

But where were the environmentalists who were looking at these issues from an Islamic perspective? I was familiar with the writings of Seyyed Hossein Nasr as a philosopher, but I didn't quite grasp the nature of his approach to the environment until I attended one of his lectures in London in the mid-1980s. Then there was Hafiz Basheer Ahmad Masri, who pioneered animal welfare from an Islamic perspective. There were also Muslims like the late Egyptian, Mostafa Tolba, who was the Executive Director of the UN Environment Programme (UNEP) and the Malaysian S. M. Mohamed Idris, President of Sahabat Alam (Friends of the Earth) Malaysia, and no doubt numerous others, but their approach was strictly secular. It was not as if environmentalism and concern for animal welfare were absent in the Islamic discourse, as one would discover in a reading of the *Island of Animals* written in the twelfth century, but these issues were not receiving anything like the kind of attention they deserved given the looming crisis (see 'Producing Results'. below).

Getting to grips with this I found myself back at university again in my late fifties discovering how the Qur'an dealt with the natural world, only to discover that Islam in its very essence was a faith deeply rooted in the natural world and it had something unique to offer humanity in search for solutions to a self-induced catastrophe. I have spent my time before and after formal university learning in the 'University of Life', in all its multi-faceted faculties, learning from people at all levels, especially those willing to get their hands dirty.

The Muslim contribution to this alliance for human survival can be profound. Islam has in its teachings both an ethical perspective and a method of practical application that can begin to meet the challenges that we face. The Qur'an is inherently environmental, and the basis of its eco-theology is now being used to educate people in many parts of the world. Islamic jurisprudence contains a body of practical applications which can provide positive solutions, and it has been proven to change attitudes almost overnight – as in the case of the fishermen in Zanzibar.[21] To multiply these activities to the levels necessary to produce lasting and successful solutions is a priority Muslims should now be addressing. I shall be devoting the rest of this chapter to the contribution Islamic teachings can make to the success of this endeavour.

Rediscovering the Natural World

A holistic approach

Today more than half the world's population live in cities. Our increasingly urban lives have ensured an almost total disconnection from the natural world to the extent that it is not uncommon to meet children who believe that milk originates in plastic supermarket cartons. It is as well to be conscious of this as much is left out in their education, both religious and secular, to the extent that we are continuing to cause irreparable harm to the Earth. Muslims are no exception to this, having also adopted urban lifestyles and modern modes of living that reveal an almost complete lack of awareness as to where this is all going. Taking children to the zoo or visiting national parks is more in the nature of recreational activity and actually reinforces the idea of otherness. The educational aspects remain lamentably superficial, and the spiritual is nowhere to be seen.

In the Islamic order, care for the natural world expresses itself in every aspect of personal behaviour. The guidance for this comes from the Qur'an and Prophet Muhammad's teachings (Sunnah). It is integral to life, an expression of existence in submission to the will of the Creator in harmony with the cosmic pattern. There is a code of conduct that governs interpersonal behaviour and an individual's rights and responsibilities within a community; it also deals with an individual's behaviour towards other sentient beings and the rest of the natural world. As Muslim interaction with the environment evolved, it manifested itself in a range of rules and institutions, embodying a truly holistic expression of life. It took into consideration the importance of reducing waste, being abstemious, generous, considerate, moderate, unselfish, caring, and sharing. The qualities of love, humility, trust and justice remained preeminent. Over the centuries these values have been elaborated upon by a succession of mystics, scholars, jurists and teachers responding to real problems experienced by the growing community of Muslims in various parts of the world.

However, this Islamic mode of expression is now severely attenuated, having been swept aside by the forces of history – like the other older traditions – into a domain which treats the natural world exclusively as an exploitable resource. As the secular ethic progressively seeped into the Muslim psyche, and as industrial development, economic indicators and consumerism became the governing parameters of society, there has been a corresponding erosion of the Muslim perception of the holistic and a withering of its understanding of the sacred nexus between the human community and the rest of the natural order.[22]

As the Islamic tapestry unfolded in its expression over the centuries, we discover that there are no references in the Qur'an to the environment or its -isms, as we now understand them. The context in which we discuss this subject today originated as recently as the second half of the last century and Rachel Carson is recognized as having given impetus to this movement since her research was published in 1962 (see chapter 2, 'A Delayed Reaction').[23] Neither can the word 'nature', which is an abstraction, be found in the Qur'an. The closest term in modern Arabic usage is *bi'ah*, which connotes a habitat or a surrounding. Nevertheless, the Qur'an is inherently environmental and holistic in its approach. It speaks of creation (*khalq*) and contains over 250 verses where this word is used in its various grammatical forms. Derived from the root *kh–l–q*, it is used in many ways to describe what we see, feel and sense in the world. The verses contain references to the natural world from herbs to trees, from fish to fowl, to the sun, stars and skies.[24] *He (Allah) created everything and determined them in exact proportions* (25: 2). The human community is but an infinitesimal part of the natural world but we have now lost sight of this through our proclivity for dominating it. The Qur'an observes: *The creation of the heavens and the earth is far greater than the creation of humankind. But most of humankind do not know it* (40: 57). Creation is the fabric into which the tapestry of life is worked.[25]

An Islamic approach to the environment provides us with a three-faceted response:

- The first is to approach the Earth as sacred and identify how through abusing it we have desacralized our understanding of it. Hence we can rediscover what it is that we have lost in the process.
- The second is to formulate an ethical understanding of our relationship with the natural world and build a knowledge base that enables us to

resolve the immediate crisis and motivates us to lay down a basis for long term behavioural change.

- The third is to look at this through the prism of political economy and discover what it is that an Islamic approach to this subject can do to both improve our understanding of the current malaise and provide some answers as to how we can create a model of wellbeing without having to subvert our own existence.

Sachiko Murata and William Chittick have an interesting approach in that they deploy traditional Islamic learning to teach modern courses on Islam. They chose as their model 'a famous and authentic hadith [Prophetic teaching] that Muslim thinkers [and teachers] have often employed for similar purposes in classical texts.'[26] Specifically, they cite a hadith popularly known as the Hadith of Jibril (Gabriel) narrated by Umar ibn al Khattab, the second caliph to succeed the Prophet after his death. This was the occasion when a stranger dressed in white sat by the knees of the Prophet, laid his hands on his thighs and began to question him about Islam. When the stranger had left the Prophet told his Companions: 'He was [the Archangel] Gabriel. He came to teach you your religion'. The Arabic word the Prophet used for religion is *dīn*[27] (sometimes spelt *deen*) and the answers to the stranger's four questions contain the basis of the *dīn* of Islam. The questions were about submission, faith, doing what is good or beautiful and about the last day.

The word 'Islam' is understood by two integral components. The first is the name attributed to the faith as a whole and the second as submission, that is submission to the will of the Creator, meaning living in harmony with the natural order. The Hadith of Jibril shows that *islam* (submission) in its verbal form is one part of a tripartite model of expression alongside *īmān* (faith) and *ihsān* (doing what is good or beautiful). Leaving aside the Prophet's responses to the fourth question, the other three responses can be viewed as a three-dimensional geometric image (see below) in which each dimension is considered independently but understood as an integrated whole.[28] In taking advantage of this method, these terms can be further defined to help us understand this matter more clearly in relation to our current-day dilemmas. I have attributed the sacred to *islam*, ethics to *īmān* and political economy to *ihsān*. Each of these dimensions influences the other two in many profound ways. In emphasizing the integrated nature of this model it is understood that sacredness is all-pervasive in that all of creation is sacred; ethics is not an exclusive domain as it is about maintaining human relationships and defining how we relate to other sentient beings and the natural world; and political economy sits comfortably with this as it is about how we conduct our affairs in the name of a just order.

The *dīn* of Islam is about the natural order and *islam*, *īmān* and *ihsān* are firmly embedded in its matrix. What follows is an overview to this approach.

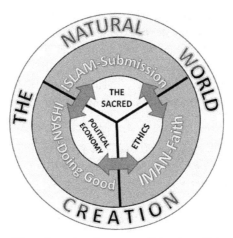

The tripartite elements of the *dīn* of Islam

The sacred[29]

There is a well-established tradition in Islam that the whole Earth is a place of prayer: A sacred space where one might contemplate the Divine. The daily activities we carry out in this space require exemplary behaviour as our every act is expected to be in the nature of a prayer. Prayer and the natural world are irrevocably connected. The term in the Qur'an used to describe the verses it contains is *āyah* (plural *āyāt*), which means signs. This term is also applied to everything in the natural world, as in:

> *There are signs on the Earth for people with certainty.*
> *And in your selves as well. Do you not then see? (51: 20–21).*

This would allow me to say that abusing the natural world and one's self is not a great deal worse than doing the same to the Qur'an itself.[30] As Nasr observes:

> Muslim sages who referred to the cosmic or ontological Qur'an ... saw upon the face of every creature letters and words from the pages of the cosmic Qur'an. ... They remained fully aware of the fact that the Qur'an refers to phenomena of nature and events within the soul of man as *ayat* (literally signs or symbols) a term that is used for the verses in the Qur'an. ... For them (the sages) the forms of nature were literally *Ayat Allah*.[31]

These verses also urge us to consider the integration of the self with the cosmos. In times gone by, when human beings did not differentiate between the self and the natural world nature was integrated into the human psyche. There was no separate environmentalism, intellectualism, capitalism, consumerism and all the other -isms that drive wedges between ourselves and the natural world. We now need to regain that consciousness we once had; an awareness that we are deeply and irrevocably interwoven into its fabric and a realization that by causing it grievous bodily harm we harm ourselves. Time now for a reappraisal, to gain a fresh understanding of what the sources, the Qur'an and the Sunnah, tell us about creation, our place in the natural order and the responsibilities we have to shoulder.[32] The following verses from the Qur'an tell us more about the signs of God:

Allah sends down water from the sky,
and by it brings the dead earth back to life.
There is certainly a Sign in that for people who hear.

There is instruction for you in cattle:
From the contents of their bellies,
from between dung and blood,
We give you pure milk to drink,
easy for drinkers to swallow.

And from the fruit of the date palm and the grapevine
you derive both intoxicants and wholesome provision.
There is certainly a Sign in that for people who use their intellect.

Your Lord revealed to the bees [saying]:
'Build dwellings in the mountains and the trees,
and also in the structures which men erect.
Then eat from every kind of fruit
And travel the paths of your Lord,
Which have been made easy for you to follow.'

From inside them comes a drink of varying colours,
Containing healing for mankind.
There is certainly a Sign in that for people who reflect (16: 65–69).

There are signs on the Earth for people who hear, use their intellect and those who reflect.

The Qur'an describes creation in numerous ways, and the first revelation[33] came in the following form to Prophet Muhammed: *Read! In the name of your Lord who created; Created man from clots of blood* (96: 1–2). In recounting this episode, Özdemir reminds us that the Prophet responded by saying he didn't know how to read; tradition has it that he was not literate. There was also no text to read, but this signifies a completely different way of looking at the world. 'The key notion is that this reading should be in the name of our Sustainer'[34] who gives existence and meaning to everything else. This reading is from the texts, *āyāt* (signs) that make up the natural world. All life emerges from the Creator and all natural phenomena are to be viewed as if they were from the book of the Creator. The revealed Qur'an signposts us to the ontological Qur'an thus:[35]

<u>The cosmos:</u>

It is He who appointed the sun to give radiance; and the moon to give light, assigning it in phases ... in the alteration of night and day and what Allah has created in the heavens and earth there are signs for people who have awareness (10: 5–6).

<u>Animals from water:</u>

Allah created every animal from water. Some of them creep on their bellies, some that walk on two legs and some on four. Allah creates whatever He wills (24: 45).

<u>Water:</u>

And We send down water from the sky and make every generous species grow in it (31: 10).

<u>Plants and crops:</u>

It is He who produces gardens, both cultivated and wild, and palm trees and crops of diverse kinds (6: 141).

On almost every page one turns to in the Qur'an there is some kind of reference to the natural world. Its approach is holistic, and it deals with the human in the natural world as integral to it. However, this has been taken for granted and there is now a disconnection between ourselves and nature to the extent that we treat it as the other – separate from ourselves. This gap is more acute now than it has ever been and we now see the natural world as a resource to be exploited. We once took from it for our survival, but now we exploit it for our pleasure and aggrandizement. Paradoxically, the gap widens as we come to know and understand more about natural phenomena, from subatomic particles to the distant galaxies.

The following verse in the Qur'an gives us a succinct description of our place in creation:

So set your face firmly towards your dīn
as a pure natural believer
Allah originated you in His original creation.
There is no changing Allah's creation.
That is the true dīn,
but most people do not know it (30: 30).

The human species emerged from the womb of the natural world at a favourable period in its vast history. The Earth rotates on a finely defined axis at an optimal distance from the Sun. The ozone layer protects us from the Sun's dangerous cancer-inducing radiation. The air allows us to breathe in a finely-balanced mix of gases, while photosynthesis process produces the oxygen we cannot live without. The right proportion of carbon dioxide in the atmosphere gives us equitable weather. The tropical rainforests act as massive air conditioners. The rhythms of the seasons are in place. The moon's orbit regulates the tides. In the temperate zones mountains store winter snows and their springtime meltwater feeds rivers that irrigate the pastures. In warmer zones the rain cycle feeds rivers copiously, enabling us to plant crops. There are fish in the sea, fowl in the glades and sheep in the meadows while bees pollinate our next harvests of grain and fruit as they buzz away their brief lives.

The primordial condition of the Earth into which humankind was brought into being is described as the *fitrah* in the Qur'an. Some translators describe this as the 'natural pattern', others the 'original state' or 'pattern', and yet others simply as 'nature'. Other scholars describe *fitrah* as 'pure state' or 'state of infinite goodness', the conscious expression of which rests uniquely with humankind. It is said that the human at birth is in a state of *fitrah*, unspoilt and pure. The term *fitrah* is derived from the Arabic root *f-t-r* and occurs once in the Qur'an. It appears in its verb form, *fatara*, fourteen times. *Fitrah* is a feminine noun which allows us to consider nature as mother. As the translators grapple to transmit the meaning of this verse, there is a simplicity inherent in this message that conveys to us a sense of where we belong in the pattern of Allah's creation. Try as it might, the human species cannot change the natural world. It would be like a car mechanic attempting to modify a nuclear reactor, but the scale is beyond comparison. Yes, we have modified the environment to our temporary advantage, as is our wont, but the human race has now exceeded its limits. This verse taken together with the rest of the verses on creation lays down the foundation for the deep ecological insights of the Qur'an. An appreciation of this would lead us to address the urgent environmental concerns of today at their root. There is a legacy owed to our succeeding generations – let it be one that earns their gratitude rather than their loathing.

This could also take us to interesting places when we explore the terms that emerge out of the root *f-t-r*[36] that appear in the Qur'an. For example:

Futur – a rent, fissure, flaw as in, '*return thy gaze; seest thou any fissure?*' (67: 3).
Tafattara – to be rent asunder as in, '*the heavens well nigh are rent above them*' (42: 5).
Infatara – to split open, be cloven asunder as in, '*when heaven is split open*' (82: 1).

The following verse from the Qur'an is said at the beginning of the standing position in the first cycle of the five daily prayers: '*I have turned my face to Him who originated (fatara) the heavens and the Earth, a pure natural believer*' (6: 79). The Qur'an is telling us that creation occurs as a seed splits to produce plant life, or like the sperm that splits the egg to produce sentient life forms. When the Qur'an says that Allah originated the heavens and the earth, could it be telling us that the universe as we know it came into existence when the heavens were rent asunder? The big bang perhaps? This is a matter for reflection.[37]

Ethics

Following our integrated model, *īmān* (faith) was described by the Prophet Muhammad as knowing with the heart, vocalizing with the tongue and acting with the body. This requires the profession of faith in God and acceptance that life in all its expressions emanates from the Divine source. The difference between *islam* and *īmān* is expressed in the Qur'an thus: '*the desert Arabs say, "we have faith", tell them: "you do not have faith, rather say, 'we have submitted', for faith has not yet entered your hearts"*' (49: 14). Submission is not the same as faith as one can submit outwardly without knowing anything inwardly. Faith grows in the heart through knowing and inspirational knowledge, it is the inner conviction that provides the foundation for the conduct of Islamic society. The Qur'an asserts, '*We did not create heaven and earth and everything between them as a game*' (21: 16). This spells out the human relationship with the natural world. As Ismail and Lois Faruqui explain:

> All Muslims therefore agree that nature was meant to be used for a moral end. It was not created in vain or sport (as a game) but as the theatre and means for moral striving. ... [I]t was made by God both good and beautiful to the end of serving man and enabling him to do good deeds. Its [nature's] goodness is derived from that of the divine purpose. For the Muslim, nature is *ni'mah*, a blessed gift of God's bounty, granted to man to use and enjoy, to transform in any way with the aim of achieving ethical value. ... Nature enjoys in the Muslim's eye a tremendous dignity.[38]

As observed earlier we now dwell in a house of cards. In spite of its laudable objectives, the 2015 Paris Agreement on climate change is another prop that is shoring up this rickety dwelling. Climate change scientists predict a disastrous feedback loop, but this has already been present in a financial system that feeds all the other loops. Short of hoping that we can all change to a more equitable way of conducting our affairs, we are left, as the Faruqis put it, with transforming our lives in the hope of 'achieving ethical value'.

Consumerism has become so all-pervasive that we even consume religion in the form of ritual. If this ritual is lifeless and does not connect us with the 'blessed gift' of the Creator, the natural world, we are left with an entity we see as a resource that satiates our consumer lives. What we are shoring up is a model of progress and prosperity that lures us like ... a *mirage in a spacious plain which a man athirst thinks it is water but when he reaches it, he finds it to be nothing* ... (24: 39).

So how do we reconnect with the Earth even as we tear it apart? Traditionally we were all unconscious environmentalists long before this word came into fashion, but having lost this connection we now desperately need to regain it.

'*Allah gives each thing its created form and then guides it*' (20: 50). The basis of Islamic social action is to establish the good and prohibit the bad. '*Let there be a community among you that calls for what is good, urges what is right and forbids what is wrong; those are the ones who have success*' (3: 104). This establishes a caring relationship with the natural world. Conservation in Islam is associated with good behaviour, which is the principle by which Muslims are expected to conduct their affairs and manage their surroundings: '*Eat of their fruits when they bear fruit and pay their due on the day of their harvest. And do not be wasteful. For He (Allah) does not love the wasters*' (6: 141).

The human species is the primary beneficiary of the Creator's handiwork. The Qur'an asserts '*be thankful*' (45: 12) as '*He has subjected all that is in the heavens and the earth for your benefit as a gift from Him*' (45: 13). But Creation is not a playground as '*We did not create heaven and earth and everything between them as a game*' (21: 16). There was a purpose in creation since '*We did not create the heaven and the earth and everything between them to no purpose*' (38: 27). This purpose is to test believers, '*He (Allah) wanted to test you regarding what has come to you. So compete with each other in doing good*' (5: 48).

What we now refer to as Islamic environmentalism was a natural way of life when Muslims lived in a way that emulated the example of the Prophet. 'The character of the Messenger of Allah was the Qur'an'.[39] The verse on waste above (6: 41) demonstrate how he exemplified this. The Prophet discouraged any wasteful action amongst his Companions, as this hadith shows:

> When the messenger of Allah passed by Sa'd as he was performing his ablutions he said, 'What it is this extravagance?' Sa'd asked, 'Can there be any extravagance even in ablution?' The messenger of Allah replied, 'Yes, even if you are on the bank of a flowing river.'[40]

The Qur'an emphasizes the value of trees and other vegetation, ... *and the plants and trees prostrate themselves (to the Creator)* (55: 6), and the lesson is driven home by the Prophet in the following hadith:

> There is none amongst the Muslims who plants a tree or sows seeds, and then a bird, or a person or an animal eats from it, but is regarded as a charitable gift for him.'[41]

Regarding the treatment of animals, the Qur'an states: '*There is no creature crawling on the earth, or those that fly, who are not communities like yourself*' (6: 38). The following hadith is a narration about an incident near a well:

> A man was walking on a road when he became very thirsty. He found a well and went into it and drank and came out. There was a dog panting and eating earth out of thirst. The man said, 'This dog has become as thirsty as I was.' He went down into the well and filled his shoe and then held it in his mouth until he climbed out and gave the dog water

to drink. Allah thanked him for it and forgave him. [The Companions asked the Prophet]: 'Messenger of Allah, do we have a reward for taking care of beasts?' He said, 'There is a reward for everyone in serving all living beings'.[42]

These examples demonstrate the depth of the material and remind us of the need to once again reconnect with the natural world. The task is not just one of reexamining the material, but to present it in such a manner that it can be readily understood, absorbed, and applied. I will deal with this in some depth in the section below under the heading 'Producing Results'.

Political economy

Ihsān is the third part of our three-dimensional model. It is described as the act of worshipping Allah as if you see Him, knowing that even if you do not see Him, He sees you. This goes far beyond ritual prayers, as every good action is also seen as an act of worship. This is commonly expressed by Muslims as 'doing what is pleasing to Allah', the Ever-Present. The discharge of our responsibilities should be tempered by justice and kindness with the intention always to do good; the ultimate reward is goodness itself. *'Will the reward for doing good be anything other than good?'* (55: 60). The end result is to create a just order, to *'uphold justice and bear witness to Allah, even if it is against yourselves, your parents, or your relatives. Whether they are rich or poor'* (4: 135). Hashim Dockrat explains:

> In Islam the complete existential social contract, which is binding on the individual within society, is contained within the idea of *jama'at* (congregation). This binding of the complete social contract is based on *ibada* (ritual, faith and worship), *mu'amalat* (public and private affairs, the zone of commercial and related transactions), and *imara* (political governance).[43]

Certain exemplary practices, like decision-making through consultation (*shūra*), were established by the Prophet and one of the primary functions of Islamic governance is to ensure the fair distribution of wealth. These were the beginnings in Islam of what we call today 'political economy', which emerged with the collection and distribution of the zakat tax, the third pillar of Islam: *'Zakat is for the poor, the destitute, those who collect it, reconciling people's hearts, freeing slaves, those in debt, spending in the way of Allah, and travellers. It is ordained by Allah. Allah is All-Knowing, All-Wise'* (9: 60).

Collectors are appointed, and paid a stipend. Those that are eligible to give are identified and their means to contribute, whether in kind or gold and silver, are estimated. Those that are needy are identified as are their levels of need. What is collected is then distributed. This all requires some form of administration, particularly if the scale is beyond what a small community can cope with. This requires leadership and in this way governance enters society. The rule of law and administration follow, trade is facilitated, and markets are established. No space is left for the accumulation of wealth within which oligarchs can breed. As Islam expanded, these institutions grew in sophistication and more were added, such as the administrative courts (*hisbah*), charitable foundations (*awqāf*), the treasury (*bayt al-māl*) – but emphatically no banks.

And in this regard we meet money again and how Islam regards it. Usury is totally forbidden in Islam, with a vehemence to it that is unmatched anywhere else in the Qur'an; be it consuming alcohol, gambling or fornication: *Those who take ribā* (usury/interest) *will rise up on the Day of Resurrection like someone tormented by Satan's touch* (2: 275); and *'forego any remaining usury if you are a true believer. If you do not know that it is war from Allah and His Messenger'* (2: 278–279). We have seen in previous chapters how the regime of the fractional reserve banks has contributed to the environmental debacle we face today and how this is linked to the political system that governs all our lives. The point is that if we do not deal with this issue we will lose the battle to recover the planet from the grievous harm we are causing it, as the global finance and banking industry drives the economic model that is bringing the Earth to its knees.

Yasin Dutton asks: 'Why, then, might this issue of *ribā* be so important to our subject matter of Islam and the environment?' to which he responds that 'it is the economic activities of present-day humankind that are the main cause of the environmental destruction that we see happening around us.'[44] As the world is ordered today there is not a great deal any Muslim country – and there are nearly sixty of them – can do about this. Not even the Organisation of Islamic Cooperation (OIC); not even Saudi Arabia or the other Middle Eastern oil and gas producing states, in spite of their massive reserves of oil, and this demonstrates the extent to which the banks have captured the entire human community.[45] We are all held captives inside the ravenous belly of the banking system.

Although much has been written about Islam, money and economics, Shaykh Abdalqadir as-Sufi is notable in having spelt out in 1985 prescient reflections about the dangers modern banking poses:

> Is the imminent collapse of the biosphere's linked ecological systems not the dominant crisis? ... Is the enslavement of the global population in a debt-spiral from which a monetary system suggests no exit not the underlying crisis from which all others stem? ... Politics is powerless ... even the super state groupings and inter-state forums, have no access or control over the financial system which in effect governs the world.[46]

JK Galbraith, the leading American economist and diplomat in the second half of the last century, said that, 'The process by which banks create money is so simple that the mind is repelled'.[47] John Gray, Emeritus professor of European Thought at the London School of Economics opines, 'Global capital markets ... make social democracy unviable.'[48] It would seem that in spite of these observations by leading thinkers the heavy human footprint sinks deeper into the Earth by the day and Joel Kovel, the American scholar and author, gets closer to the core of the matter in the preface to the second edition of his book, *The Enemy of Nature*:

> *The Enemy of Nature* ... tries to give expression to an emerging and still incomplete realization that our all-conquering capitalist system of production, the greatest and proudest of all the modalities of transforming nature which the human species has yet devised, the defining influence in modern culture and the organizer of the modern state, is at heart the enemy of nature and therefore humanity's executioner as well.[49]

In spite of these observations, which appear to be growing by the day, it is something of a puzzle that there has not been equally trenchant and considered responses from the droves of Muslim scholars and economists who populate these areas of concern. Could it be that in their endeavours to create an Islamic version of modernity they are missing the point? There is, however, an Islamic version of the good life and if it is to be defined in materialistic terms this can be articulated from the Qur'an, for Islam is not anti-materialist: *'It is He (Allah) who created everything on the Earth for you'* (2: 29). And the Qur'an duly warns against excess: *'Do not forbid the good things Allah has made lawful to you. And do not over step the limits'* (5: 87).

So, in Islamic terms how do we set about having a reasonably satisfactory lifestyle and at the same time deal with global warming? In a sense it is back to the future and here is an outline of the Islamic principles dealing with trade, finance, environmental protection and governance we could be examining:[50]

- Equitable development based on the principle of establishing a just society is an imperative. The concept of *iqtisād*[51] encourages moderation and simplicity; material benefits are not denied and one group or individual cannot take undue advantage over others in the distribution of *limited* resources.
- Hoarding and the accumulation of wealth for its own sake is discouraged. The Qur'an asserts in the chapter entitled 'Competition', *'Fierce competition for this world distracted you until you went down to the graves. No indeed you will soon know! Again no indeed you will soon know!'* (102: 1–4)
- Distribution and sharing are encouraged. The institutions used for this purpose are zakat which is 2½ per cent of savings and is compulsory; *Sadaqah* is the voluntary giving of surplus wealth; *al-Infaq* is an extension of *sadaqah* whereby the deprived are seen as having a right over the surplus wealth of others: *'Have you seen him who denies his faith? He is the one who harshly rebuffs the orphans and does not urge the feeding of the poor'* (107: 1–3).
- Caring for the environment is encouraged: *Corruption has appeared in the land and sea, for that men's own hands have earned, that He (Allah) may let*

them taste some part of that which they have done, that perhaps they may return (30: 41).

- Money, broadly, is any non-perishable commodity that can be used as a medium of exchange. Gold and silver have historically been the precious metals used for minting coinage. Usury (bank interest) is forbidden and in Islam nothing like the current fractional reserve banking system and the issue of paper money is possible.
- Markets based on Islamic teachings must be free. Prices are determined by open transactions and produce must be open to inspection. Hoarding, monopoly trading, gazumping, false or misleading information are forbidden.
- Some examples of contractual obligations are: *Shirkah*, a partnership in which the lender shares the risk; *Mudārabah*, an agreement between the provider of capital and labour; *Bay' al-Salam*, an advance payment based on the weight of produce and delivery time (futures trading of the variety practised today are prohibited); *Qard hasan*, a beautiful loan (no interest is charged and there is no time limit; the onus of repayment is on the borrower and who is honour-bound to repay at the first opportunity).[52]
- Wages: 'Give the worker his wages before his sweat dries'.[53]
- Consultation: '*Conduct [your] affairs by mutual consultation*' (42: 38).
- Justice: '*Be upholders of justice, bearing witness for Allah alone, even against yourselves or your parents and relatives, whether they are rich or poor*' (4: 135).

Producing Results

Exploring the legacy

Numbering approximately 1.5 billion people, the Muslims are estimated to constitute 20 per cent of the world's population and they have something quite unique to offer in resolving the dilemmas we now face. If their leadership, both scholars and politicians, would take on their share of responsibility in this area of concern then much could be achieved in helping to resolve the environmental debacle we face today. There is also a view that this leadership, if it emerges at all, will be too late to make a difference, so the solution must lie in each one of us taking responsibility.

The relearning processes needs to be swift and the signs of the times we live in require a dissemination of the teachings in a form that is appropriately identifiable, can be quickly assimilated and then produce practical solutions in situations that require change for the better. This can work in two stages, and as observed earlier the first is to assemble a body of knowledge from the Qur'an that deals with the natural world, which could be referred to as Knowledge of Creation (*'Ilm'ul-Khalq*). The second is to rediscover the rules of Environmental Governance (*Fiqh al-bi'ah*) which has evolved over the centuries and can be alternatively referred to as Islamic Natural Resources Management.

Although a lifelong lover of nature, I did not really begin to look for a specifically Islamic narrative of the natural world until the mid-1980s, which was when terms like 'environmentalism' and 'ecology' began to assume prominence. At a meeting of conservationists, I could only respond inadequately to a question for an explanation of the Islamic approach to conservation. There were people who motivated me such as Mostafa Tolba, the Egyptian head of the United Nations Environment Programme (UNEP) from 1975 to 1992, and SM Idris of Malaysia, who was a recipient of the alternative Nobel Prize called the Right Livelihood Award in 1988. However, it proved almost impossible to find Islamic scholars with any knowledge of this subject relating to current issues; a clear sign of the retreat of Islamic scholarship into a narrow self-defeating literalism and the sucking of the body politic of Muslims into the quicksands of globalization. But there were precious gems to be mined.

The late al-Hafiz BA Masri's well researched book *Animals in Islam*[54] was published in 1989, and although frail of health when I met him in 1990 he was full of enthusiasm for his chosen subject. This was my first personal encounter with someone who was both a writer and an activist from an Islamic standpoint. I then discovered *The Island of Animals*,[55] the translation of a tenth-century Arabic manuscript of a fable concerning human and animal relationships. The research of Mawil Izzi Dien had led him to the discovery of tenth and fifteenth century Islamic scholars writing about animals, their habitats and their relationship with humans.[56] Seyyed Hossein Nasr has had a long standing reputation in the academic world as a Muslim thinker and writer who wrote for a western audience. I discovered his *Man and Nature*,[57] which has universal appeal and incredibly was first published in 1968. I had some catching up to do studying his encyclopaedic works, and he was always available for advice and support when I needed it.

In my quest for solutions I belatedly discovered Environmental Protection in Islam[58] in the mid-1990s. The first edition was jointly published in 1983 by the Meteorology and Environmental Protection Administration (MEPA) of Saudi Arabia and The World Conservation Union (IUCN). Although it provided me with answers to the kinds of questions I was asking it still remains a reference work for specialists, despite the brave efforts of the likes of Othman Llewellyn.[59]

The poetry of the Sufis[60] had much to say about the natural world. The compositions of Jalal ud-Din Rumi,[61] which are popular amongst western audiences, are perhaps the best known amongst them. There are others like Shaykh Muhammad Ibn Al Habib[62] of Morocco, less well known even amongst Muslims, who eulogize nature in their poems. The following is a section from 'Reflections', one of the *qasidas* (poems) from the *Diwān*[63] (a collection of poems) of the Shaykh:

> Reflect upon the beauty of the way in which both the land and sea are made,
>   and contemplate the attributes of Allah outwardly and secretly.
> The greatest evidence to the limitless perfection of Allah can be found

Both deep within the self and the distant horizon.[64]
If you were to reflect on physical bodies and their marvellous forms
And how they are arranged with great precision, like a string of pearls;
And if you were to reflect on the earth and the diversity of its plants
　　and the great varieties of smooth and rugged land in it;
And if you were to reflect on the secrets of the oceans and their fish,
　　and their endless waves held back by an unconquerable barrier;
And if you were to reflect on the secrets of many winds
　　and how they bring the mist, fog and clouds which release the rain;

And if you were to reflect on all the secrets the heavens –
　　the Throne and the Foot-stool and the spirit sent by the command –
Then you would accept the reality of *tawhīd*[65] with all your being,
　　and you would turn from illusions, uncertainty and otherness;

The literature I was able to access at that time, although sparse, was encouraging and provided clues regarding the depth of this subject. There was the hint of the existence of an Islamic environmentalism that needed to be expressed in a way that cut through the confusion of our times: 'We were in a pickle of paradigms and the essence of the Islamic worldview was lost in the acid of the dominant secular ideology'.[66] It was at this point that I decided to research this myself and opted to enrol for a master's degree in Islamic Studies at the University of Birmingham in 1990, only to discover that Islam was inherently environmental in scope and it was a belief and value system deeply embedded in the natural order. During my research I was hardly able to find a page in the Qur'an that did not mention the cosmos, or the earth, or the sentient beings that thrived in it, and numerous other aspects of the natural world. It was a puzzle that Muslims have allowed this knowledge to lapse, and unravelling this should throw some light not only on how and why we were detached from these teachings but also on what caused it. The preceding chapters of this book represent the unravelling that took me to that point.

The following year I resigned from my position in the Commission for Racial Equality[67] after twenty-three years' service. The choices in front of me were to continue working behind a desk or to move on to something that was fresh and dynamic and open to exploration. The second option was compelling. I had the good fortune to have Ibrahim Surty[68] as my lecturer on the subject of the Qur'an, who had a deep and profound knowledge of the subject. When he discovered that I had enrolled to understand the approach of the Qur'an towards the environment he said: 'Do you know Allah describes His creation in the Qur'an as His *āyāt* (signs)?' This was the kind of encouragement I needed.

The challenge was to drive home to Muslims environmental lessons in the Qur'an in a way that could be easily presented, absorbed and acted upon. Whilst I was putting this material together I had the good fortune to meet Dr Abdullah

Omar Naseef in Jeddah in 1998 and discuss my project with him.[69] He was at the time serving as the Vice-President of the Consultative Assembly (*Majlis ash-Shūra*) of Saudi Arabia. A dedicated environmentalist himself, he warmed to the idea and we came to the conclusion that this body of knowledge should be aptly called *'Ilm'ul-Khalq* (Knowledge of Creation). The outcome of this was a teaching resource called 'Qur'an, Creation and Conservation'.[70]

This resource was successfully developed under the aegis of the Islamic Foundation for Ecology and Environmental Sciences (IFEES), which I founded in 1994. Apart from being presented at various international fora, they have also been used as core messages in the development of IFEES' training resources, which are now being used internationally from West Africa to Indonesia. The effectiveness of these techniques can be demonstrated by their success in Zanzibar, where it was first used in 1999 as a pilot. The target audience were fishermen who had resorted to the use of dynamite for their fishing activities. This practice stopped immediately after the workshops and the fisherman consequently participated in a conservation programme to save fish stocks. A teaching resource[71] was produced at the request of the local community and this project was also made the subject of a documentary produced by an American media company.[72]

'Ilm'ul-Khalq (Knowledge of Creation)

The principles of *'Ilm'ul-Khalq* are derived directly from the Qur'an. The Qur'an is not a textbook, and we look to certain principles in it that communicate the environmental message in a way that could both be understood and acted upon as a spur to change. The following is a crystallization of what I consider to be the essentials that will bring into focus the changes needed today from an Islamic perspective. The planetary system, the Earth and its ecosystems, all work within their own limits and tolerances. Likewise, Islamic teaching sets limits to human behaviour as a control against excess and it can be said that the limits to the human condition are set within four principles: The unity principle (*tawhīd*), the creation principle (*fitrah*), the balance principle (*mīzān*), and the responsibility principle (*khilāfah*). The following is a brief look at these principles under their respective headings:

Tawhīd: The Unity principle

Tawhīd is the foundation of Islamic monotheism and its essence is contained in the declaration (*shahādah*) that every Muslim makes and it is a constant reminder of faith. It runs 'There is no god but Allah' (*lā ilaha illā'llah*)[73] and is the foundational statement of the unity of the Creator from which everything else flows. The Qur'an instructs us to *'Say: "He is Allah, the One, Allah the Eternal"'* (112: 1). It is the testimony to the unity of all creation and to the fabric of the natural order of which humankind is an intrinsic part: *'What is in the heavens and the earth belongs to Allah: He encompasses everything'*

(4: 126). This is the bedrock of the holistic approach in Islam as this affirms the interconnectedness of the natural order.

Fitrah: The Creation principle

Fitrah is the principle that describes the primordial nature of creation: 'Allah originated you in His original creation' (30: 30). Humankind was created within the natural pattern of nature and, being of it, its role is defined by that patterning. The term fitrah was discussed earlier in this chapter in the section under 'The Sacred'.

Mīzān: The Balance principle

In one of its oft-recited passages the Qur'an describes creation thus:

> The All-Merciful taught the Qur'an,
> He created man, and taught him clear expression.
> The sun and moon both run with precision,
> the plants (stars) and the trees all bow down in prostration,
> He erected heaven and established the balance ...
> so that you would not exceed the balance (55: 1–8).

Allah has singled out humankind and taught it clear expression; given it the capacity to reason. If the sun, the moon, the stars did not run with precision or the trees serve their purpose, by submitting to Allah, it would be impossible for life to function on earth. Our actions have disrupted the balance (mīzān) of creation. We have managed to change the climate, melt the glaciers at the poles and on mountaintops, poison rivers, drain lakes, level mountains, corrode the coral in the oceans, poison the soil, denude the forests and cause the extinction of other species. It doesn't seem to have entered our collective consciousness that we could be even a threat to ourselves. This is short-lived in terms of cosmic time and as latecomers we have behaved outrageously. We are the only sentient beings in creation who, through the very gift of reasoning, can choose not to submit. That is refuse to live in harmony with the rest of creation and destroy everything around us by our presumed ingenuity. And 'He created man from a drop of sperm and yet he is an open challenger' (16: 4). This geological epoch is now coming to be known as the anthropocene: the human species has now itself become a force of nature.[74]

Khilāfah: The Responsibility principle

Khilāfah is the name given to someone who deputizes for or stands in for someone else.[75] This is usually substituted by terms such as steward, guardian, successor, and other similar expressions: 'It is He (Allah) who appointed you as stewards on the Earth' (6: 165). The Creator 'offered the trust to the heavens, the earth and the mountains, but they refused to take it on and shrank from it. But man took it on.

He is indeed wrong doing and ignorant' (33: 72). The enormity of this trust, which imposes a moral responsibility on humankind, is expressed metaphorically: there is a seeming paradox in the heavens and the earth and the mountains refusing to undertake it. The humblest of God's creation – *The creation of the heavens and the earth is far greater than the creation of humankind* (40: 57) – has taken on the weightiest of all responsibilities. This is the price we pay for the gift of intelligence: *'He (Allah) created man and taught him clear expression'* (55: 3–4), which is the exclusive privilege of the human race; giving our species the capacity to communicate and bring changes to the natural world at will.[76]

'Adl (justice) is the basis with which we are required to execute this trust: *'weigh with justice and skimp not in the balance. He set the earth down for all beings. With its fruits, its palm trees with clustered sheaths'* (55: 9–11). The role of *khalifah* is thus a sacred duty handed down to the human race, as no other sentient being in creation can perform this role. We are required to care for and manage the Earth in a way that conforms to God's intention in creation: it should be used for our benefit without causing damage to the other inhabitants of planet Earth who are communities like ourselves; *'There is no creature ... who are not communities like yourself'* (6: 38). The relationship we have over the natural world is not a right to do as we please but a responsibility which carries with it the burden of accountability.

We can deduce in outline from these four principles that creation is complex and finite. It emerged from one source and was designed to function as a whole. Like the rest of creation, humankind is made in a state of goodness with the potential for good actions. It is inextricably part of this pattern of creation, but it is the only element of it that can choose to act against the Divine will by using the very gift of reasoning bestowed upon it by the Creator. Submission to the Divine will, the natural law that holds in check the instincts of man the predator, is the way to uphold our responsibilities as the Creator's *khalifa*. In other words, humankind is the guardian of the natural order.

This route leads us to identifying the core values that can be easily taught and understood, and the following four contain the essentials for the messages that we need to put across:

- Love (*Hubb*):

  'It is He (Allah) who originates and regenerates (nature). He is Ever Forgiving, All-Loving' (85: 13–14).

- Humility (*hawn*):

  'The servants of the Compassionate (Allah) are those who walk the Earth in humility' (25: 63).

- Trust (*Amānah*):

  'We (Allah) offered the trust to the heavens, the Earth and the mountains; they refused to take it and were afraid of it. But man took it on. He is indeed wrong doing and ignorant' (33: 72).

- Justice (*'Adl*):

> *'We (Allah) sent our messengers with clear signs. And (We) sent down the Book and the Balance with them so that mankind might establish justice'* (57: 25).

Fiqh al-Bi'ah (Jurisprudence of the Environment)

The Qur'an is a manual for life and a source-book for good behaviour, not the least of which is our relationship with the environment. The Prophet of Islam is sometimes referred to as the Qur'an walking, because by what he said and did (his Sunnah) he interpreted the Qur'an for us. As we know he left no aspect of life untouched and he said and did much to conserve and protect the environment. The Qur'an and the Sunnah form the basis of the Shariah and, as Islam spread, scholars interpreted this in response to the needs of the people from which emerged the jurisprudence (*fiqh*) of Islam. Although there is no separate jurisprudence on the environment as such, there exists within the body of *fiqh* much that regulates our behaviour to promote a conservationist approach. This body of knowledge is now beginning to be recognized as *fiqh al-bi'ah* (Islamic environmental jurisprudence), the revival of which, to meet current challenges, is being pioneered by Indonesian scholars.[77] At the same time the pioneering work of stalwarts like Othman Llewellyn in Saudi Arabia also needs to be acknowledged. I shall now take a cursory look at some of the conservation techniques Muslims used in the past that have been lost to us because of our corrupted relationship with the natural world. These include the *himā*, now being revived in certain parts of the Middle East, *harīm*, for its potential, *awqāf*, for its universality, and the *hisbah*, which provides the all-important element of oversight coupled with sanctions.

The *Himā* system

Himā means a place that is guarded or one that is forbidden to enter. In practice, a *himā* could broadly be defined as a reserve or a conservation zone. It is pre-Islamic in origin, when the practice was for powerful tribal leaders to reserve pasture and grazing land for their exclusive use, allowing limited access to others. The Prophet abolished these practices when he established the first *himā* in Islam announcing: 'There are no reserves except for Allah and His Messenger.'[78] His first reserve was created on land adjacent to the hill of al Baqi' (also referred to as al Naqi') near Madinah as pasture for the cavalry horses of the Ansar (the people of Madinah; literally, the 'Helpers') and the *Muhājirūn* (the people of Makkah who took refuge in Madinah; literally the 'Emigrants'). There are two scholarly opinions about this. The first interpretation takes the Prophet's words literally, holding that no further *himā*s are permitted; the second takes a more practical view, that the Prophet always set an example to be emulated. He forbade exclusivity and the examples set by the rightly-guided caliphs and successive rulers always followed this basic rule, holding that *himā* were for the people.

It is possible for a state, agency or community to establish *himā*s as long as they are utilized for the 'public good, such as conservation and management of rangeland, forest and woodlands, watersheds and wildlife'.[79] There is considerable flexibility in the establishment of *himā*s as long as these core values are adhered to, which are conservation and protection. The establishment of *himā*s:

- Is enforced by a legitimate authority or community sanction.
- Incorporates the principle of public welfare.
- Strikes a balance between environmental protection and the needs of local people.
- Protects ecosystems from deliberate damage and exploitation.
- Adheres to the core values defined in the Qur'an.

As we begin to appreciate the value of an institution in Islam that has proved its worth over the centuries we could legitimately be asking some searching questions, such as why revisit *himā*s now? The answer could be that Muslims are now belatedly rediscovering their Islamic conservation heritage. This is a response to the times we now live in after a period not just of neglect but of opposition to a system thought of as being out of date. For example, the *himā* system in Saudi Arabia was decimated after the discovery of oil, real estate then taking precedence over conservation.

If the *himā* was a system legitimized by the Prophet, then why was this largely unknown in the non-Arab Muslim world? One of the reasons was that environmental degradation was not an issue in the past. It is an issue now triggered of by concerns relating to population pressure, climate change, deforestation, pollution and other familiar outcomes of human activity that negatively impacts on the environment. Another reason is that in the widely dispersed Muslim world encompassing ecosystems and climates that bore no relation to each other, people had developed culturally compatible systems for themselves. These widely dispersed local systems bear a remarkable similarity to the *himā* system we have been discussing. In the example of Zanzibar given previously, it was the conservation-conscious local people who initiated the project, although they were unaware of the principles underlying the *himā* model.

What of government policy? This is something to be reckoned with because governments are primarily development-oriented, which means the relegation of conservation issues are lower down the order of priorities. Faith-based initiatives are considered irrelevant or archaic as they supposedly do not bring in the revenue. Governments can also be influenced by large-scale institutional funding, from the World Bank for instance, for outsize projects that prioritize economic growth over all else. For example, the Zanzibar project has suffered from this kind of intervention. There is a good case here for making environmental ethics education compulsory for politicians before they take office.

Every community that was interested in its survival, no matter what part of the world they lived in, had a method and a system that they had developed over long periods of time to protect and preserve their land, e.g., the Indonesians call it the *Adat* system, some Moroccans, the *Agdal* system, and the Omani system of canal conservation is called *Aflaj*. The *himā* is now assuming generic proportions, and whether it is in Indonesia, Zanzibar or Morocco the same principles apply and local people can chose to call it by the name they are most familiar with.[80]

Himās are human-integrated, natural resource-management systems and have traditionally been associated with the following five types:

- Areas where grazing of domestic animals is prohibited.
- Areas where grazing is restricted to certain seasons.
- Beekeeping reserves where grazing is restricted during flowering.
- Forest areas where cutting of trees is forbidden.
- Reserves managed for the welfare of a particular village, town or tribe.

In practical terms, most things are possible so long as the core values are adhered to.

The *Harīm* system

As we have seen, the *himā* system lends itself to a range of arrangements, and its inherent flexibility allowed it to function sustainably long before the concept itself was invented. Within the matrix of the Shariah, there are also specialized systems that may or may not lend themselves to a *himā*-type approach but could equally well address sustainability issues.

There are two considerations that apply to the *himā* model that apply equally well to the *harīm* system. The first is that old systems have been hastily discarded in preference to so-called modern methods or superseded by a development mentality that disregards the old ways, which are only then belatedly rediscovered as being relevant and appropriate to local conditions. The other consideration is that the models in the *fiqh* invariably mirror traditional models in most parts of the world, after having made allowances for local conditions. As discussed earlier, this proved to be the case with the *himā* system as well. What this demonstrates is the reality of our primordial nature (*fitrah*). No matter how distant people are from each other and no matter how different local habitats and ecosystems are from each other, people's responses to protect their spaces have remarkable similarities. This could be described in one word, and that is survival. These survival responses have been developed over millennia and as we spurn them in the name of progress and development we invite the unintended consequences that follow: degraded soil, landslides, mudslides, degradation of water resources, overgrazing, the consequences of using excessive amounts of toxic chemical fertilizers and insecticides, etc.

The value of the Shariah is that it has codified traditional methods and thus preserved working models to serve as examples of good practice for future generations. This does not entail spurning scientific innovation, but it does mean being choosy about new ideas that increasingly make claims upon our attention, the principle behind this being the sustaining of land use by good conservation practices. To give an example, the indiscriminate use of chemical fertilizers has poisoned agricultural land, groundwater and adjacent watercourses in many parts of the world, thus depriving people of land and sources of water that were once productive. Old tried and tested systems were discarded for what was peddled to local people as modern scientific methods. The gain was short term, the pain long term. These processes are now thankfully being re-examined, though not soon enough.

Now largely in disuse, the *harīm* system designates various spaces as inviolable or sacred zones within Islamic law. These zones are public utilities or natural resources that are protected from being corrupted or changed to other uses for which they were not originally intended. The Prophet prohibited pollution and misuse in three specific places of public resort and these spaces became part of the *harīm* or the inviolable zone. The following locations could be designated as *harīm*:

- Sources of water such as rivers, springs, wells, watercourses, lakes and seas: For example, in the *fiqh* literature, the *harīm* for a river covers half the width of the river on both banks; for a well, a radius of forty cubits[81] from the centre of the well with variations depending on local conditions; and for a spring there are variations depending on local conditions.[82]
- Spaces surrounding a town or village that provide renewable resources such as forage and firewood: A tree providing shade considered to be of public benefit can be part of a *harīm* and its zone of protection can extend to a radius of five cubits or more.
- Utilities, such as roads and public squares.

Whilst the *himā* system lends itself to considerable flexibility, the *harīm* system is much less so because of the concept of sanctity reflected in its name and the possibility of legal sanctions for violating its sanctity. It is no accident that the two holiest places in Islam, Makkah and Madinah, are known as *al-Haramayn* (from the same root as *harīm*). These places are sanctuaries for people, wildlife and vegetation and on the day the Prophet conquered Makkah he declared: 'It is sacred by virtue of the sanctity conferred on it by God until the Last Day. Its thorn trees shall not be cut down, its animals shall not be disturbed, the objects lost within it shall be picked up only by the one who will announce them, and its fresh herbage will not be cut.'[83] With regard to the last prohibition there is an exception, as at the request of one of his Companions the Prophet allowed the harvesting of a cooking herb called *al-idkhir*.[84] The Prophet made a similar declaration in Madinah, designating the area between the mountains and lava flows surrounding the city as sacred. He also made the surrounding area of Madinah to the extent of twelve miles, or nineteen kilometres, a *himā* type reserve. The distinction here is that game cannot be hunted in a *harīm* zone whilst it is allowed in the *himā*. This appears to define the difference between the two systems: there is total prohibition in the *harīm* whilst there is flexibility in the *himā*.

Apart from the *Haramayn*, there are no places that can be identified today as *harīm* or prohibited zones. With the advent of the modern nation-state, the old tried-and-tested systems are almost all extinct and the closest analogues currently are the state-managed national parks. These are by nature large spaces that are mostly out of the control of local people. However, with the state and local people working together there are possibilities that could be explored to allow for the reintroduction of the *harīm* system in eco-sensitive areas in the Muslim world.

There are huge pressures on the guardians of the *Haramayn* today imposed by the vastly increased number of pilgrims that visit Makkah and Madinah each year, not the least of which are the 'Dangerously high levels of air pollutants'[85] that are being released during the pilgrimage. There is also much debate about whether the right balance has been struck between providing the appropriate amenities for visiting pilgrims and honouring the Prophets' Sunnah. There is now a groundswell of feeling against what is happening in the two holy places, as those who are responsible for maintaining them are seen as having surrendered to excessive commercialization and insensitive modernization.

Awqāf

Awqāf (singular *waqf*), is an Islamic institution that encourages individuals to set up charitable endowments through gifts and bequests for the public good. It originated when Umar ibn al-Khattab, the second caliph of Islam, asked the Prophet what he should do with some land he had been given. The Prophet suggested that he make the land an endowment and give its produce in charity. Umar's son reported that his father gave the property to charity declaring that: 'It must not be sold or gifted or inherited and that the harvest would be devoted to the poor, kinsfolk, the freeing of slaves, for the cause of God, for travellers and for guests.'[86]

Commonly, a *waqf* is a trust comprised of land or buildings dedicated in perpetuity for charitable purposes prescribed by the donor. It could also be a financial trust, a business or a share in the proceeds of a business. The specific purposes of a trust can vary according to the wishes of the donor, and the income derived from a trust could be used for conservation purposes such as protecting forests, wildlife, watercourses and sensitive habitats. A *himā* or a *harīm* can also be endowed in perpetuity and their specific purposes defined by the donor. For example, a *himā* could be established as a *waqf* comprising of land for use by a named community or village or neighbourhood for specific agricultural and or grazing purposes, similar to the Agdal pastoral community in Morocco.[87] Or a *harīm* in a forest area for the protection of wildlife, like the orangutan in Indonesia or the rhinoceros in Africa, could be dedicated in perpetuity as a *waqf*. The scope of the *awqāf* system can be gleaned from the fact that there was a *waqf* for stray dogs in Makkah, although whether this *waqf* is still in existence today is uncertain.

Hisbah

The term *hisbah* refers to the realization of an ethical vision based on verse 3: 104 of the Qur'an: *Let there be a community among you that calls for what is good, urges what is right and forbids what is wrong; those are the ones who have success.* This is a call to good action addressed to the individual, but as Islam expanded and governance became more complex the need to widen the scope of this guidance took the form of an administrative court system called the *hisbah*, which also had powers of enforcement. The *hisbah* is administered by an office holder known as the *muhtasib* with powers to make immediate judgements based on local custom and practice (*'urf wa-'ādah*) and to enforce them. The first *muhtasib* appointed by Umar, the second caliph of Islam, was a woman called Shifa' bint Abdallah. The *hisbah* functions as an adjunct to the formal Islamic courts and has the advantage of immediacy and enforceability. The following are some of the differences between the powers of a *muhtasib* and an individual or community in giving expression to verse 3: 104 of the Qur'an.[88]

Muhtasib	Individual/Community
Acts with authority of the state.	Acts on the basis of local concern.
Cannot transfer authority of the office.	Unrestricted right to act on basis of concern.
Obliged to respond to complaints.	Not obliged to respond to complaints.
Has powers of investigation.	Has no powers of investigation.
Can employ assistants.	Cannot employ assistants for this purpose.
Can impose sanctions.	Cannot impose sanctions.
Receives a salary from public funds.	Acts voluntarily.
Can make legal rules based on local custom and practice (although not on the Shariah).	Cannot make legal rulings based on local custom and practice.

The Shariah in perspective

There has been much loose talk about the Shariah, and a brief word here regarding its value, specifically in relation to conservation and its allied subjects, would be germane at this point. Its literal meaning is 'road' and defined as, 'the legal modality of a people based on the Revelation of their Prophet'.[89] However, 'If the Shariah can be described as a vast carpet with intricate patterns woven into it, what I have done here is to borrow some of these patterns from the complex weave of the carpet and make sense out of them'.[90] I have dealt with the relationship between the Qur'an and the Sunnah in the previous section, especially in regard to what can be done immediately by applying these teachings, but the picture will not be complete if I do not deal with the background to this, however briefly.

I glean the following broad code in the Shariah as it relates to land management and the natural world:[91]

- The elements that compose the natural world are common property.
- The right to benefit from natural resources is a right held in common.
- There shall be no damage or infliction of damage bearing in mind future users and the effect this will have on biodiversity.

As an extension of this foundational code, Muslim legalists have over the centuries established the following principles. A person invalidates his rights over a particular natural resource if by exercising it they:

- Cause detriment to another; and we can add to this, cause detriment to biodiversity.
- Cause detriment to another without corresponding benefit to the other and the region's biodiversity.
- Cause general detriment to society by degrading ecosystems.

In addition:

- Every member of society is entitled to benefit from a common resource to the extent of his or her need so long as he or she does not violate, infringe or obstruct the equal rights of other members of society and does not degrade local ecosystems.
- Accountability rests with the user.
- In return for benefits derived from a renewable resource the user is obliged to maintain its value.

If the user causes destruction, impairment or degradation he or she is held liable to the extent of putting right the damage caused.

Legislative Principles:

- Allah is the sole owner of the earth and everything in it. People hold land on usufruct – that is for its utility value only. There is a restricted right to public property.
- Abuse of rights is prohibited and penalized.
- There are rights to the benefits derived from natural resources held in common.
- Scarce resource utilization is controlled.
- The common welfare is protected.
- Benefits are protected and detriments are either reduced or eliminated.

Institutions

In addition to the institutions discussed under *'Ilm'ul-Khalq* and *fiqh al-bi'ah* above, the Shariah also defines the following institutions supporting equitable land use:

(a) People who reclaim or revive land (*ihyā' al-mawāt*) have a right to its usufruct.

(b) Land grants (*iqtā'*) may be made by the state for reclamation and development.

(c) Land may be leased (*ijārah*) for its usufruct by the state for its reclamation and development.

A way forward

The development and application of these principles and institutions have seen if not their complete obliteration, a steep decline in recent centuries, as another world view that saw the natural world exclusively as real estate and a source of profit gradually took hold. We are experiencing the consequences of this now. However, there are clear indications here as to how this Islamic heritage could again be put to good use for the welfare of the entire planet and all life on it.

Muslims have their own unique part to play in finding solutions to our current dilemmas, but we also need to consider that in today's global order – of which Muslims are a significant part – economic growth as a path to conspicuous consumption tops the list of priorities. Muslim nation states,[92] of which there are now about sixty, are ardent adherents to this consumer ethic. It should be obvious from this that it becomes almost impossible for Muslims, whether individuals or nation states, to live today according to a normative Islam. There is now a schizoid tendency in Muslim society whereby it strives to maintain its deep attachment to Islam while it insists on enjoying the fruits of what we call modernity.

One could say with a reasonable degree of certainty that the environmental problems we encounter today would not have arisen in a society ordered in accord with Islamic principles because its world view 'defined limits to human behaviour and contained excess'.[93] Safeguarding against human excess has the effect of protecting the natural world. Human behaviour is governed by the Shariah, which evolved holistically, and there is nothing to stop its further evolution to address contemporary issues. However, there are important impediments to its proper application today in what is now a complex political climate. They are:

- The Shariah is marginal even in so-called Islamic states because of the dominance of the global system now in place. The influence of international trade and finance is a case in point, not to mention the global cultural assault made possible by accelerating communications technology.
- The *hisbah* is an agency that has the potential to set down environmental guidelines and act to resolve conflict in matters relating to natural resource utilization, but it is now virtually extinct.
- The nation state and the apparatus of government have separated themselves from the body of Islamic scholars (*ulema*) who are coming to be known as 'the religious authorities', a euphemism for a clergy, which is not recognized in Islam.[94]
- Following the dominant Western model, the specialists and ministries of Muslim states increasingly function in watertight compartments. As a mirror of what is happening in the West, Muslim economists and environmentalists tend to be two separate species with opposing perspectives.
- The nation state model, which is now the universally accepted form of governance that all Muslim countries have adopted, considers economic development based on the untenable growth model as its highest priority. Coping with issues relating to the environment is much lower down the scale.

Whilst there are impediments to the implementation of solutions based on Islamic principles it is important for Muslims to both engage in the debate concerning the environment and at the same time work in partnership with the other traditions and like-minded groups and organizations, if only for the fact that future unborn generations have a common inheritance; we all share the same Earth. We will discover that there are more than willing partners out there, and if Muslims were true to themselves their spontaneous inclination would be to prioritize the welfare of others with whom they share a finite planet.

If anything, this crisis should remind Muslims of the deep ecology of their faith – that is the unified and holistic nature of creation – and all life forms are integral to it. Climate change makes no distinction between race, religion, culture and lines on the map. As Muslims comprise a fifth of the world's population, potentially they should at least be contributing a fifth part towards the solution. It is now for the scholars to produce the motivation, the mystics the inspiration, the politicians to give a direction and the rest of us to make sure it happens.

Endnotes

1. Fazlun Khalid, 'Guardians of the Natural Order', *Our Planet*, 8/2 (July 1996). *Our Planet* is the journal of the United Nations Environmental Programme (UNEP).
2. Carroll Quigley, *The Evolution of Civilizations: An Introduction to Historical Analysis* (Carmel, IN: Liberty Fund, 1979), p. 367.
3. John Grim and Mary Evelyn Tucker, *Ecology and Religion* (Washington, DC: Island Press, 2014), p. 1.
4. Seyyed Hossein Nasr, *Man and Nature*.
5. Thomas Berry, *The Great Work: Our Way into the Future* (New York: Bell Tower/ Random House, 1999).
6. Abdullah Omar Naseef, 'The Muslim Declaration on Nature – Assisi 1986' (ARC, undated). Available at: http://www.arcworld.org/faiths.asp?pageID=132 (accessed on 8 May 2018).
7. Fazlun Khalid and Joanne O'Brien (eds), *Islam and Ecology* (London: Cassell, 1992).
8. 'The Ohito Declaration on religions, land and conservation, March 1995' (ARC, undated). Available at: http://www.arcworld.org/news.asp?pageID=871. See also, 'The Ohito declaration on religions, land and conservation' (IFEES, undated). Available at: http://www.ifees.org.uk/ohito-declaration/ (both accessed on 29 May 2018).
9. The MOA (http://www.moainternational.or.jp/en/intro/intro1.html) is a Japanese spiritual movement that is committed to the conservation ethic. Alternative link: http://www.moa-fresno.org/about-moa.html (accessed on 29 May 2018).
10. Fazlun M Khalid, 'Applying Islamic Environmental Ethics', in Richard Foltz (ed.), *Environmentalism in the Muslim World* (New York: Nova Science, 2005), p. 93.
11. 'Sacred Gifts: What is a Sacred Gift?' (ARC, undated). Available at: http://www.arcworld.org/projects.asp?projectID=49 (accessed on 8 May 2018).
12. 'Tanzania: Fishermen say no to dynamite – using Islamic environmental principles' (ARC, January 2011). Available at: http://www.arcworld.org/projects.asp?projectID=170; 'Zanzibar' (IFEES, 23 April 2015). Available at: http://www.ifees.org.uk/zanzibar/ (accessed on 8 May 2018).
13. 'Saudi Arabia: Rare species protected in first national biosphere reserve' (ARC News, undated). Available at: http://www.arcworld.org/projects.asp?projectID=173; OA Llewellyn, M Hall, AG Miller and T M al-Abbasi, 'Important Plant Areas in the Arabian Peninsula: 4. Jabal Aja', *Edinburgh Journal of Botany*, 68/2 (July 2011), pp. 199–224. Available at: http://journals.cambridge.org/action/displayAbstract?fromPage=online&aid=8292828 (accessed on 10 May 2018).
14. Mary Evelyn Tucker and John Grim, who are now based at Yale University, USA, are possibly the world's leading academics in this field of study. They run the Forum on Religion and Ecology (FORE; http://fore.yale.edu).
15. Richard C Foltz, Frederick M Denny and Azizan Baharuddin (eds), *Islam and Ecology: A Bestowed Trust* (Cambridge, MA: Harvard University Press, 2003).
16. Press Release, 'Religions for the Earth: A Multifaith Gathering to Respond to Climate Change at Union From September 19–21, 2014' (Religions for the Earth, 15 July 2014). Available at: http://ipcc.ch/report/ar5/docs/Religions_for_the_Earth.pdf (accessed on 8 May 2018).
17. GreenFaith: Interfaith partnership for the environment (http://www.greenfaith.org).

18. Larry Rasmussen, Declaration for an Awakened Kinship with the Earth (Center for Earth Ethics, undated). Available at: https://centerforearthethics.org/resources/declaration-for-an-awakened-kinship-with-the-earth/ (accessed on 8 May 2018).

19. 'Encyclical Letter *Laudato Si*' of the Holy Father Francis on Care for Our Common Home' (Libreria Editrice Vaticana, undated). Available at: http://w2.vatican.va/content/francesco/en/encyclicals/documents/papa-francesco_20150524_enciclica-laudato-si.html (accessed on 8 May 2018).

20. 'Islamic Declaration on Global Climate Change' (IFEES, undated). Available at: http://www.ifees.org.uk/declaration/ (accessed on 8 May 2018).

21. See 'Zanzibar' (IFEES, 23 April 2015).

22. Khalid, 'Islam and the Environment', p. 332.

23. Carson, *Silent Spring*.

24. Khalid, 'Islam and the Environment', note 23, p. 334.

25. Op cit. note 23, p. 335.

26. Sachiko Murata and William C Chittick, *The Vision of Islam: The Foundations of Muslim Faith and Practice* (London: IB Tauris, 1996), p. xxv.

27. 'It has often been observed that Islam cannot ordinarily be described as a "religion" and that it prescribes a way of life that goes beyond the performance of rituals. The word used in the Qur'an for religion is *dīn*, and it appears in this context in over seventy different places, often in circumstances that place it outside the domain of ritual. *Dīn* in essence describes an integrated code of behaviour that deals with personal hygiene at one end of the spectrum to our relationships with the natural world at the other. It provides a holistic approach to existence, it does not differentiate between the sacred and the secular and neither does it place a distinction between the world of man and the world of nature.' This quote first appeared in Khalid, 'Islam and the Environment', p. 332. Besides the seventy references, there are twenty other references to *dīn* that alludes to a particular day, namely *Yawm al-Dīn* (the Day of Reckoning).

28. Murata and Chittick, *The Vision of Islam*, pp. xxxii–xxxiv.

29. The term 'sacred' can be understood both as something deserving veneration, as in God, or an entity that needs to be treated with great respect, as in the Earth. The term is used here in the second sense. For the Oxford Dictionary of English's definition see https://en.oxforddictionaries.com/definition/sacred (accessed on 9 May 2018).

30. Fazlun Khalid, 'Exploring Environmental Ethics in Islam: Insights from the Qur'an and the Practice of Prophet Muhammad' in John Hart (ed.), *Religion and Ecology* (Oxford: Wiley Blackwell, 2017), p. 133.

31. Seyyed Hossein Nasr, *The Need for a Sacred Science* (Richmond: Curzon Press, 1993), pp. 130, 131.

32. Khalid, op cit. note 366, p. 133.

33. Arberry observes: 'These revelations were supernaturally received, in circumstances of a trance-like nature, over a considerable number of years intermittently'. See Arthur J Arberry, *The Koran Interpreted* (Oxford: Oxford University Press, 1964). In the Islamic tradition this event is described as the appearance of Archangel Gabriel.

34. Ibrahim Özdemir, 'Towards an Understanding of Environmental Ethics from a Qur'anic Perspective', in Foltz, Denny and Baharuddin, *Islam and Ecology*, p. 7.

35. Khalid, op cit. note 366, p. 134.

36. Hanna E Kassis, *A Concordance of the Qur'an* (Oakland, CA: University of California Press, 1983), p. 451.

37. Khalid, op cit. note 366, pp. 131, 132, 133.

38. Faruqi and Faruqi, *The Cultural Atlas of Islam*, p. 322.

39. *Sahih Muslim*, Book 6, Hadith 168 (Sunnah.com, undated). Available at: https://sunnah.com/muslim/6/168 (accessed on 9 May 2018).

40. *Sunan Ibn Majah*, Book 1, Hadith 460 (Sunnah.com, undated). Available at: https://sunnah.com/urn/1254240 (accessed on 9 May 2018).

41. *Sahih Bukhari*, Book 41, Hadith 1 (Sunnah.com, undated). Available at: https://sunnah.com/bukhari/41/1 (accessed on 9 May 2018).

42. *Muwatta Malik*, Book 49, Hadith 1696 (Sunnah.com, undated). Available at: https://sunnah.com/urn/417560 (accessed on 9 May 2018).

43. Hashim Ismail Dockrat, 'Islam, Muslim Society and Environmental Concerns: A Development Model Based on Islam's Organic Society', in Foltz, Denny and Baharuddin, *Islam and Ecology*, p. 345.

44. Yasin Dutton, 'The Environmental Crisis of Our Time: A Muslim Response', in Foltz, Denny and Baharuddin, *Islam and Ecology*, p. 331.

45. Riyad Asvat, *Economic Justice and Shari'a in the Islamic State* (Melbourne: Madinah Press, 2011), p. 149.

46. Shaykh Abdalqadir as-Sufi, *The World Crisis* (Unpublished pamphlet, Murabitun, Norwich, 1985).

47. Quoted as an epigraph in Anthony Sampson, *The Money Lenders* (London: Hodder and Stoughton, 1981), p. 29.

48. John Gray, *False Dawn: The Delusions of Global Capitalism* (London: Granta, 2009), p. 88.

49. Joel Kovel, *The Enemy of Nature: The End of Capitalism or the End of the World?* (London: Zed Books, 2007), p. vii.

50. Fazlun Khalid, 'The Environment and Sustainability: An Islamic Perspective', in Colin Bell, Jonathan Chaplin and Robert White (eds), *Living Lightly, Living Faithfully: Religious faiths and the future of sustainability* (Cambridge: Faraday Institute of Science and Religion, 2013), pp. 235–237.

51. *Iqtisād* deals with Islamic thoughts on finance, banking, economics, trade, commerce and industry.

52. Khalid, 'The Environment and Sustainability', pp. 65, 72, 97–112, 136, 151.

53. Sunan Ibn Majah, Book 16, Hadith 2537 (Sunnah.com, undated). Available at: https://sunnah.com/urn/1267610 (accessed on 9 May 2018).

54. Al-Hafiz BA Masri, *Animals in Islam* (London: Athene Trust, 1989).

55. Denys Johnson-Davies (trans.), *The Island of Animals* (London: Quartet Books, 1994).

56. Mawil Izzi Dien, *The Environmental Dimensions of Islam* (Cambridge: The Lutterworth Press, 2000), pp. 25, 26.

57. Seyyed Hossein Nasr, *Man and Nature*, see note 4.

58. Abubakr Ahmed Bagader, Abdullatif Tawfik El-Chirazi El-Sabbagh, Mohammad As-Sayyid Al-Glayand, Mawil Yousuf Izzi-Deen Samarrai, and Othman Abd-ar-Rahman Llewellyn (eds), 'Environmental Protection in Islam', Environmental Policy Law Paper No. 20, second revised edition (IUCN, Gland Switzerland

and Cambridge UK, 1994). Available at http://cmsdata.iucn.org/downloads/eplp_020reven.pdf (accessed on 9 May 2018).

59. Othman Llewellyn is an American born environmentalist who has dedicated his life to reviving the historic conservation models used in Islamic societies of the past. He lives and works in Saudi Arabia and has been responsible for pioneering the establishment and maintenance of the Jabal Aja' Biodiversity Reserve in that country, which serves as a benchmark for Muslim environmentalists in other parts of the world. See 'Sacred Gifts of Saudi Arabia' (*EcoIslam*, April 2008). Available at: http://www.ifees.org.uk/wp-content/uploads/2015/04/newsletter_EcoIslam4.pdf (accessed on 9 May 2018). Also see note 15.

60. The Sufis are the mystics of Islam. Sufism is also known as *tasawwuf.* Jalal al-Din Rumi (d.1273) is perhaps the best known Sufi master in the West. He founded the Mevlevi Sufi order, famous for its whirling dervishes.

61. Rumi's best known work is *The Mathnawi*, a six volume epic didactic poem considered to be a masterpiece on the teachings of Sufism.

62. Shaykh ibn al-Habib, a distinguished scholar of Islamic jurisprudence, was also the last great Sufi Master of the Darqawi-Qadiri order in Morocco. He was 110 years old when he died on his way to Makkah to perform the Hajj pilgrimage in 1972.

63. The *Dīwān* of Shaykh Muhammad ibn al-Habib (Cape Town: Madinah Press, 2001), p. 69. This *Dīwān* was written over a period of forty years in the last century.

64. This is a reference to verses 51: 20–21 in the Qur'an, which has been the inspiration for the title of this book.

65. The doctrine of divine unity, otherwise described as Islamic monotheism.

66. Khalid, 'Applying Islamic Environmental Ethics', pp. 88–91.

67. The predecessor organization of the Equalities and Human Rights Commission in the UK.

68. Ibrahim Surty is the founder of the Qur'anic Arabic Foundation (QAF). See Muhammad Ibrahim HI Surty, *Towards Understanding Qur'anic Arabic* (Birmingham: QAF, 1993).

69. See note 6.

70. Khalid, 'Qur'an, Creation and Conservation'.

71. Fazlun Khalid and Ali Khamis Thani, *A Teachers Guide Book for Islamic Environmental Education* (Birmingham: IFEES, 2007). Available at: http://www.ifees.org.uk/wp-content/uploads/2015/04/13751866541.pdf (accessed on 9 May 2018).

72. Chedd Angier, 'Saving the Ocean with Carl Safina. Episode 9: Sacred Island' (PBS, 17 January 2013). Available at: http://chedd-angier.com/savingtheocean/Season1/Episode9.html (accessed on 9 May 2018).

73. The second half of this declaration is 'Muhammad is the Messenger of Allah' (*Muhammad ar-Rasul Allah*).

74. Khalid, op cit. note 31, p. 132.

75. Aisha Bewley, *The Glossary of Islamic Terms* (London: Ta-Ha, 1988), p. 12.

76. Khalid, op cit. note 31, p. 138.

77. Ahsin Sakho Muhammad, Husein Mumammad and Roghib Mabrur et al. (eds), *Fiqih Lingkungan* (Jakarta: Laporan, 2004). This is the first text of its kind to appear in this form anywhere and sets a benchmark for the evolution of *fiqh al-bi'ah*. There are plans for the work to be translated into English.

78. Bagader, op cit. note 59, pp. 42–45.

79. Op cit. note 59, p. 25.

80. MK Suleiman, et al. (eds), 'International Workshop: Towards an Implementation Strategy for the Human Integrated Management Approach Governance System: Theories, Concepts, Methodologies, Case Studies and Action Plans' (Kuwait Institute for Scientific Research, 2013). Available at: http://hima.kisr.edu.kw/main/assets/publications/HIMAProceedings.pdf (accessed on 9 May 2018).

81. A cubit is approximately 22 inches (43 centimetres).

82. Bagader, op cit. note 59, pp. 35–37.

83. Op cit. note 59, p. 26.

84. *Sunan An Nasa'i*, the 'Book of Hajj', 2892 (Sunnah.com, undated). Available at: https://sunnah.com/urn/1129020 (accessed on 9 May 2018).

85. Adapted Media Release, 'Severe air pollution spikes during yearly pilgrimage to Mecca' (*Medical News Today*, 20 December, 2014). Available at: http://www.medicalnewstoday.com/releases/287062.php?tw (accessed on 9 May 2018).

86. Bagader, op cit. note 59, p. 27.

87. Suleiman, op cit. note 82, p. 51.

88. Abu'l Hassan al-Mawardi and Assadullah Yate (trans.), *Al-Ahkam As Sultaniyyah: The Laws of Islamic Governance* (London: Ta-Ha, 1996), pp. 337–362.

89. Bewley, *The Glossary of Islamic Terms*, p. 21.

90. Khalid and Thani, op cit. note 73, p. 7.

91. This is culled from an unpublished paper I produced in the mid-1990s for training purposes. My sources for this summary were:

    a) Bagader, Abubakr Ahmed, Abdullatif Tawfik El-Chirazi El-Sabbagh, Mohammad As-Sayyid Al-Glayand, Mawil Yousuf Izzi-Deen Samarrai, and Othman Abd-ar-Rahman Llewellyn (eds), 'Environmental Protection in Islam', Environmental Policy Law Paper No. 20, second revised edition (IUCN, Gland Switzerland and Cambridge UK, 1994). Available at http://cmsdata.iucn.org/downloads/eplp_020reven.pdf.

    b) Izzi Dien, Mawil, *The Environmental Dimensions of Islam* (Cambridge: The Lutterworth Press, 2000).

    c) Özdemir, Ibrahim, 'Towards an Understanding of Environmental Ethics from a Qur'anic Perspective', in Richard C Foltz, Frederick M Denny and Azizan Baharuddin (eds), *Islam and Ecology: A Bestowed Trust* (Cambridge, MA: Harvard University Press, 2003).

    d) Nasr, Seyyed Hossein, *Man and Nature: The Spiritual Crisis of Modern Man* (London: Unwin Paperback, 1990).

92. By my reckoning there are only two countries that have declared themselves as Islamic States and they are Saudi Arabia and Iran. The rest of the fifty-six states which are part of the Organization of Islamic Cooperation (OIC) are states with Muslim majorities run on secular lines.

93. Fazlun Khalid, 'Islamic Pathways to Ecological Sanity: An Evaluation for the New Millennium. Ecology and Development', *Journal of the Institute of Ecology*, 3 (2000), p. 7.

94. This is the case in the Sunni tradition of Islam, which accounts for about 85 per cent of the world's Muslim population. The Shia tradition has an established clergy.

6

SURVIVING
THE ANTHROPOCENE

Collapse

The human epoch

As I was beginning to write this final chapter of the book (January 2018) I received a blog from the Millennium Alliance for Humanity and the Biosphere (MAHB)[1] with the heading, 'What Will It Really Take to Avoid Collapse?'[2] Collapse refers to global Earth systems collapse and this has made my task of summarizing the book much easier. The gist of the main issues I intended to raise is included in this blog and its main link says it much better and with the authority of 15,000 scientists. However, before I proceed to discuss this I think the heading of this chapter needs a little explanation.

I devote a paragraph in the second chapter (see 'A Disconnected People') to the idea that the human race is now itself a force of nature. It would seem that the scientific community has come to the conclusion that this indeed is what we have become and have defined this age as the Anthropocene – the human epoch. We have defined this geological epoch as one that belongs to us, or more accurately as the age in which we have profoundly influenced the way the Earth works. Or in new-speak, we have tampered with the default settings of Earth systems to the extent that it threatens our own survival. However, 'the term Anthropocene was coined not by a geologist, but by the Nobel Prize-winning atmospheric chemist Paul Crutzen in 2000'.[3] This is akin to a heart surgeon telling a brain surgeon how to do her job. This is not unusual in the rarefied climate of scientific enquiry, and the problem here appears to be that Crutzen is an atmospheric chemist whilst geological time is measured by geologists. Be that as it may, the term Anthropocene has entered the lexicons and if I may crave the indulgence of the scientists who decide upon these matters I would like to suggest that we stay with the geologists, stratigraphy and the Holocene (the geological time which the proposed Anthropocene will replace) and reconsider the timing of Anthropocene as a sub-category.

A working group, reporting to a congress of geologists meeting in Cape Town in 2016, said that, 'in its considered opinion, the Anthropocene epoch began in 1950 – the start of the era of nuclear bomb tests, disposable plastics and the human population boom'.[4] Other dates such as 1800 were suggested, as this was the time when the Industrial Revolution was beginning to take a grip, but I think there is another date that needs to be considered which profoundly changed the nature of the human relationship to the Earth, and this was when usury/interest (the difference is merely academic) was legitimized by Henry VIII in 1545 (see chapter 4, 'Acceleration'). This was the event that eventually led to the creation of the Bank of England in 1694, enshrining the magical fractional reserve banking system which conjures money out of thin air and has kept the entire planetary population in its thrall ever since. In all probability, those who concocted these events had no idea of its unintended consequences and what they were getting the human race into. There is a hadith attributed to Prophet Muhammad that might explain the enormity of this event: 'The Prophet (peace be upon him) said: "A time is certainly coming to mankind when only the receiver of usury will remain, and if he does not receive it, some of its vapour will reach him". Ibn Isa said: "Some of its dust will reach him"' (narrated by AbuHurayrah).[5]

I have discussed money and banking in previous chapters (see chapter 1, 'A Lethal Cocktail' and chapter 4, 'Acceleration') and if we stop to think for a while this could quite conceivably lead us to the conclusion that the cement we have been using to build our house of cards is a creation of pure fantasy. This fantasy just about holds it together and we dream of building more layers on this collapsing edifice. Is it any wonder that we are now hurtling towards a collapse? So, it shouldn't surprise anyone that we are now fully immersed in the Anthropocene and that Crutzen and his colleagues have 'concluded that we humans have made an indelible mark on our one and only home. We have altered the Earth system qualitatively, in ways that call into question our very survival over the coming few centuries'.[6]

Heads in the sand

This brings me neatly back to the MAHB blog. To begin with it was reassuring (rather perversely) to see the image of a collapsed house of cards at the head of the blog for this was the title – 'A House of Cards' – I had chosen for the first chapter of this book; so I am not out on a limb after all. It is not only atmospheric chemists and geologists who see eye to eye on what we are doing to the planet, there are a host of other scientists who can also see the black hole that we are falling into and are now warning us for the second time in twenty-five years. The MAHB blog immediately points us to 'World Scientists' Warning to Humanity: A Second Notice'.[7] The first warning in 1992 was endorsed by more than 1,700 independent scientists, 'including the majority of living Nobel laureates in the sciences'. They cautioned humankind that, 'a great change in our stewardship of the Earth and the life on it is required, if vast human misery is to be avoided' and showed in their manifesto 'that humans were on a collision course with the natural world'. We must be grateful for this timely warning but this phraseology also tells us something about how we have come to view our relationship with the natural world. We see it as the other; a conditioned attitude of our minds. How can we be on a 'collision course' with an entity we are wholly immersed in? We have corrupted the systems that were designed to protect us and are indulging in a painfully long process of committing suicide. A picture that immediately comes to mind is that of man sawing away on the wrong side of the branch of a tree he is perched on. As he finishes his task the branch comes crashing down to Earth and he with it.

However, the scientists expressed concern about current, impending, or potential damage to planet Earth involving ozone depletion, freshwater availability, marine life depletion, ocean dead zones, forest loss, biodiversity destruction, climate change, and continued human population growth. They proclaimed that fundamental changes were urgently needed to avoid the consequences our present course would bring.[8]

'The scientists pleaded that we stabilize the human population ... (and) implored that we cut greenhouse gas (GHG) emissions and phase out fossil fuels, reduce deforestation, and reverse the trend of collapsing biodiversity.' The second notice had the unprecedented support of more than 15,000 scientists worldwide, far surpassing the first notice issued twenty-five years earlier. The only sliver of good news since 1992 is the stabilization of the ozone layer in the stratosphere, but:

> humanity has failed to make sufficient progress in generally solving these foreseen environmental challenges, and alarmingly, most of them are getting far worse ... Especially troubling is the current trajectory of potentially catastrophic climate change due to rising GHGs from burning fossil fuels, ... deforestation, ... and agricultural production – particularly from farming ruminants for meat consumption ... Moreover, we have unleashed a mass extinction event, the sixth in roughly 540 million years, wherein many current life forms could be annihilated or at least committed to extinction by the end of this century.[9]

The scientists further itemize their concerns as follows:

- intense but geographically and demographically uneven material consumption;
- continued rapid population growth as a primary driver behind many ecological and even societal threats;
- reassess the role of an economy rooted in growth;
- reduce greenhouse gases;
- incentivize renewable energy;
- protect habitats;
- restore ecosystems;
- curb pollution;
- halt defaunation;
- constrain invasive alien species.

Scientists are less prone to using emotive language than most of us but when they say: 'humanity is not taking the urgent steps needed to safeguard our imperilled biosphere' we need to sit up and take notice. They urge us to put pressure on our politicians and 'insist that their governments take immediate action as a moral imperative to current and future generations of human and other life' and compel political leaders to do the right thing. They also remind us that: 'It is also time to re-examine and change our individual behaviours, including limiting our own reproduction (ideally to replacement level at most) and drastically diminishing our *per capita* consumption of fossil fuels, meat, and other resources.'[10]

They sound a note of cautious optimism when they point to the closing gap in the ozone hole in the atmosphere and the advancements made 'in reducing extreme poverty and hunger ... rapid decline in fertility rates in many regions attributable to investments in girls' and women's education, the promising decline in the rate of deforestation in some regions, and the rapid growth in the renewable-energy sector ... but the advancement of urgently needed changes in environmental policy, human behaviour, and global inequities is still far from sufficient.'[11]

Jeremy Lent, who wrote the MAHB blog that landed on my desk, makes three very salient points.[12] First, the declaration by 15,000 scientists was largely ignored by the media. How was it possible that the greatest existential threat faced by our civilization was barely considered newsworthy? Secondly, 'The fundamental problem is brutally simple: our world system is based on the premise of perpetual growth in consumption, which puts it on a collision course with the natural world' (another collision course). Thirdly, 'the only thing that will truly advert collapse will be a radical restructuring of the economic system that is driving us ever more rapidly to that precipice'.[13]

The Qur'an observes: *Corruption has appeared in the land and sea, for that men's own hands have earned, that He (Allah) may let them taste some part of that which they have done, that perhaps they may return* (30: 41).

What Now?

Connectedness

Just as this book was about to go to press (8 October 2018) IPCC issued a Special Report on Global Warming urging policy makers to limit global warming to 1.5°C. The scientists responsible for this report urged in its press release for '... rapid, far- reaching and unprecedented changes in all aspects of society ...'.[14] This has further persuaded me to remain with the conclusions that I make in this book.

There are four themes that weave their way throughout this book. The first is the inbuilt essence of connectedness that permeates throughout planet Earth. It is fundamental to everything in existence and we can experience this in small or large measure depending on our own state of awareness. The second is the universal desire for a better life. This is a natural and legitimate aspiration, but how far we want to go down this road is another issue. For some it is just two square meals a day and a roof above their heads. For others it's two cars in the drive. This brings me to the third theme. The nature of the civilization we have created for ourselves based on perpetual growth goes against the grain of the natural world. This is a self-defeating heresy and it manifests suicidal tendencies. Fourthly, what has made all this possible and its legitimacy is the pseudo-religion of capitalism that promises us an earthly paradise whilst at the same time choking us with debt.

I attempted to show in the second chapter of this book how everything is connected and all our actions eventually lead us to taking something out of the natural world. This book came to be in your hands, the reader, by a complex process and as it wound its way to you it will not only have consumed renewable and non-renewable resources, it will also have generated a certain quantum of carbon dioxide and thus contributed its little bit to global warming. Every single thing we do uses up energy: even the mere act of thinking or sleeping. This energy originates from only one source and that is the sun. Indeed, if there was no sun there would have been no planetary system and no Earth, no photosynthesis that gives us oxygen, no warmth that creates clouds that gives us rain, no you, no me. We may also consider the fact that there would have been no life without water, which is finite and which we are busily polluting. Planet Earth has always had a finite quantity of it, recycled a myriad times and shared by all living beings over millennia. We are all irrevocably connected by the nectar of life: *We (Allah) made every living thing from water* (21: 30).

Nature functions in rhythms, in cycles. The rain cycle is the most obvious: evaporation, condensation, precipitation. Then there are the nitrogen, oxygen and carbon cycles that keep the atmosphere in balance. When I die and my people bury me the trees above me will absorb the nutrients of my decomposing body, producing new and hopefully vigorous growth. This new growth will react with the sun activating photosynthesis, producing more oxygen. It will absorb a little bit of the CO_2 overload in the atmosphere. Good to know I will be of service to my fellow beings even when I'm gone. As my nutrients reach the loftiest branches there is the chance, and also the hope, that parts of me will be communing with the stars. That is the way it has been ordained. But not so plastic and nuclear waste.

While scientists continue to warn us of an imperilled biosphere, concern for connections between the damage caused to the Earth by perpetual human conflict is conspicuous by its absence. At the same time we ravage the planet for its resources we prepare to destroy each other in the name of ideology, religion, sectarianism, nationalism, racism and any other cause we might be in thrall to at any given time. For a start, there are enough nuclear weapons in the global stockpile to wipe our civilization out many times over. It will be the shortest war in the history of humankind and there will be no survivors to celebrate victory. Additionally, 'World military expenditure was estimated at $1686 billion in 2016, equivalent to 2.2 per cent of global gross domestic product or $227 per person.'[15] Consider what this money could have done to alleviate global hunger. Consider President Putin's recent announcement that there is now a resurgence of the global arms race.[16] Here we go again – one step forward two steps backwards. Then there is the Middle East cauldron, war in Ukraine, genocide in Myanmar, ethnic cleansing in Xinjiang China, a permanent open prison in Palestine, the appalling treatment of minorities – whichever quarter they come from – xenophobia, global terrorism, insurgency, counter-insurgency; another global trade war initiated by President Trump,[17] added to which there are drug wars, the refugee crisis and much more. The UN established a Peace Building Commission in 2005, and although it is working in the now-familiar conflict areas in Africa, the Middle East and South Asia I have yet to come across any evidence to suggest that it is working towards the dismantling of the arms industry.

There is a thin dividing line between altruism and barbarism. We have a lot of growing up to do.

Search for a better life

We all want a better life. As sea routes were opening up in the sixteenth century, Europeans sailed out to new and ostensibly empty worlds in droves looking for the good life. Now this trend is reversed as people cross the Mediterranean from Africa to Europe in rubber dinghies in search of the same thing. Once the good life is experienced we appear to crave more and more of it and are characteristically loath to share our good fortune with the less fortunate. We propose building walls to separate neighbour from neighbour; erect electrified fences around refugee camps; incarcerate asylum seekers in island prisons.

We are selective in what we choose to see and hear. The bad news is beyond a distant horizon: conflict and convulsions occur in faraway places; bad weather, storms, tidal surges, floods, mudslides, drought, global warming, climate change. The cinemas are full and Hollywood and Bollywood succeed in making hay in the gloom. They whisk us away from the drudgery of our daily lives into a fantasy of excitement, adventure and romance. Likewise the television and radio; social media; and endless drugs that block surges of hopelessness and anomie. We seldom make connections between lifestyle and consequence until and unless we are directly affected. Life is now a consumer fairground. No spaces remain for the voice calling out for change. No longer time for reflection:

> *There are signs in the Earth for people with certainty.*
> *And in your selves as well. Do you not then see?* (51: 20–21)

Important life-changing news is brushed under the carpet so that hardly anyone was aware of the world scientists' 'Warning to Humanity' described as the 'greatest existential threat faced by our civilization'. Those who should take special heed of these warnings are those who have done well out of the system, while in places like Bangladesh and sub-Saharan Africa, Armageddon has already arrived and is spreading. The 'Richest 1 percent bagged 82 percent of wealth created last year (2017) [the] poorest half of humanity got nothing'.[18] A puzzling aspect of this whole affair, given its life-threatening reputation, is the indifference of the media. The 'Second Notice' of the Alliance of World Scientists did not get the coverage it deserved. To be sure there is coverage of environmental news, but not in the volume and frequency the subject demands. There is a splash during World Environment Day (5 June), as the press report on the event and hashtags appear on social media, and then the excitement subsides. My point is, environmental news should be given the same status as business and financial news – where whole pages are devoted to its coverage in the daily print media. Regular business bulletins appear on the radio and television throughout the day, so surely this subject – which is now a matter of human survival – warrants the same kind of coverage, if not more. Public awareness leaves much to be desired and the media can help mainstream this subject. So how about it?

Good news of people and organizations striving to find solutions to environmental issues are not difficult to find. In an attempt to be balanced, the scientists themselves gave as an example the rapid global decline in ozone-depleting substances. But this is the point at which complacency could be setting a trap for us. Manufacturers responsible for producing refrigerators and air conditioning appliances using ozone-depleting substances can be forced to change their practices by regulation. New production lines are also good for profit. The best route to take is to initiate positive courses of action ourselves without waiting for regulation, and this is on the increase.

Much can be done working internationally. For example, the Islamic Declaration on Global Climate Change was launched in Istanbul in August 2015 and the Arab Environmental Governance Charter emerged in December 2016.[19] National and local initiatives are vital and there are many examples of good practice, such as recycling waste, cutting down consumption of meat, cycling or using public transport as alternatives to driving cars. Trees are being planted by the billion in Pakistan: 'Cricket-star turned politician Imran Khan launched the green initiative in Khyber Pakhtunkhaw after vast areas were ravaged by floods and widespread felling.'[20] Though these responses are laudable and necessary, the nature of the world order today is so constructed that these activities, although numerous, are no more than swimming against a tidal wave. Here are just two examples of how the ideology of progress and the search for the good life affects us all.

Rural depopulation and urban drift are two sides of the same coin. This is the result of the search for a better life that drives people to cities, where more than 50 per cent of humanity now live. Vast numbers live in congested conditions which rely on a finely-balanced water supply system. What do we do when the looming water crisis finally arrives and how are we going to respond when the flush toilets stop working? At the time of writing, Cape Town in South Africa faces exactly such a dilemma.[21] It has had low rainfall for three consecutive years. There is also another side to this. As we congest in cities, who is going to grow our food? In this regard it would appear that our future lies in the hands of multinational corporations. Writing in the *Guardian* about the proposed merger between the chemical giants Monsanto, 'the world's largest seed company,' and Bayer, 'the second largest pesticide group,' John Vidal observes:

> Through its many subsidiary companies and research arms Bayer–Monsanto will have an indirect impact on every consumer and a direct one on most farmers in Britain, the EU and the US. It will effectively control nearly 60% of the world's supply of proprietary seeds, 70% of the chemicals and pesticides used to grow food, and most of the world's GM crop genetic traits, as well as much of the data about what the farmers grow where, and the yields they get.[22]

It would appear that industrial farming will fill the vacuum left by the loss of the labour force in rural areas and the supply chain will then make available processed food in plastic containers to the very people who have left their farms and moved to the cities to experience the allure of the good life. People who once ate fresh organic produce now store pre-packed dinners in their freezers with added chemical preservatives. End result: greater global warming, greater use of plastic packaging and the side effects of chemical food preservatives on overall health, which are just beginning to emerge.

The Indian government is concerned that over 100 million old people are now left alone to fend for themselves by their offspring, who have left their homes in search of the good life. This is not an exclusively Indian problem but it does tell us something about the unintended consequences of progress. India has one of the highest economic growth rates in the world, and it aspires to create an expanded consumer oriented middle class like any other self-respecting, developing nation state. To this end it has modified the education system of the colonial era, designed to create an administrative class, to a modern system that now creates a consumer class. There is a price to pay for how modernity has defined education (see chapter 1, 'A Lethal Cocktail') which acts as a springboard to the dream careers and the consumer mirage to which young people aspire. Alongside this, progress has created a problem of crisis proportions in its extremist advocacy of individualism that has progressively destroyed the ecosystem of community to which family and tradition have been closely entwined.

It came to be accepted by European intellectuals 'that history was a series of irreversible changes in only one direction – continual improvement' and marked in the eighteenth century 'by a wave of optimism about the future and the inevitability of progress in every field'[23] (see chapter 3, 'The Progress Trap'). In modern times this attitude found an echo in the words of British prime minister Harold MacMillan, when in 1957 he told the British people that they 'have never had it so good', and more recently by the American president when he promised he would 'make America great again'. What lies behind these aspirations is the creation of the good life, materially speaking, and isn't this what we all aspire to? Today the source of the good life is the bank (see reference to Norway and Denmark under *Capitalism: A debtocracy*, below), for as the banks themselves tell us, 'Without banks, we wouldn't have loans to buy a house or a car. We wouldn't have paper money to buy the things we need. We wouldn't have cash machines to roll out paper money on demand from our account.'[24]

Growth without limits?

Economic growth is the big lie. The all-consuming human race, more particularly those located in the overdeveloped parts of it, have ensured that we have exceeded the carrying capacity of planet Earth:

> Since the 1970s, humanity has been in ecological overshoot, with annual demand on resources exceeding what the Earth can regenerate each year. Today humanity uses the equivalent of 1.7 Earths to provide the resources we use and absorb our waste. This means it now takes the Earth one year and six months to regenerate what we use in a year. We use more ecological resources and services than nature can regenerate through overfishing, over harvesting forests, and emitting more carbon dioxide into the atmosphere than forests can sequester.[25]

The Qur'an warns: '*Do not strut arrogantly on the earth. You will never split the earth apart nor will you ever rival the mountains in stature*' (17: 37).

I discussed growth at some length in chapter 3 ('The Fantasy of Growth') and I will indulge in one more quote from the UK-based New Economics Foundation to drive this message home: 'indefinite global economic growth is unsustainable. Just as the laws of thermodynamics constrain the maximum efficiency of a heat engine, economic growth is constrained by the finite nature of our planet's natural resources (biocapacity)'.[26] In that case, why are the IMF and the World Bank pushing for growth? In its introduction to the *World Economic Outlook*, published in October 2017, the IMF observes:

> The global upswing in economic activity is strengthening, with global growth projected to rise to 3.6 percent in 2017 and 3.7 percent in 2018 … But the recovery [from the 2008 banking crisis] is not complete: … growth remains weak in many countries, and inflation is below target in most advanced economies … For policymakers, the welcome cyclical pickup in global activity provides an ideal window of opportunity to tackle key challenges – namely to boost potential output …[27]

In its overview to *Global Economic Prospects: Broad-Based Upturn, but for How Long?* The World Bank observes:

> The World Bank forecasts global economic growth to edge up to 3.1 percent in 2018 after a much stronger-than-expected 2017, as the recovery in investment, manufacturing, and trade continues. Growth in advanced economies is expected to moderate slightly to 2.2 percent in 2018, as central banks gradually remove their post-crisis [banking crisis] accommodation and the upturn in investment growth stabilizes. Growth in emerging market and developing economies as a whole is projected to strengthen to 4.5 percent in 2018, as activity in commodity exporters continues to recover amid firming prices.[28]

Both the IMF and the World Bank have issued statements deploring climate change, but their business is economic growth with all its consequences for global warming. How then do they square this circle? 'Green growth' is making its presence felt, but we still need to ask what is the difference between sustainable growth and sustainable development. What is the measure of the difference between the two in terms of their impact on the environment? Are the IMF and the World Bank speaking to the scientists who warn of the dire consequences of the 'business as usual' mentality? The raison d'être of the nation state is to deliver prosperity to its people with its *a priori* assumption that this can only happen by endless growth, sustainable or otherwise. This mindset has hit the buffers. We should now be looking at ways in which people can lead decent lives, possibly the very minimum of which is defined by the 2015 Sustainable Development Agenda, without causing the earth any more harm, and more particularly in how to deal with climate change with the urgency it deserves.[29] It is now time to redefine prosperity, progress and development. All these buzzwords could in my view be replaced by just one word: Equity. It could be translated as fair shares for all ensuring the just distribution of the planet's declining resources (see the next section, below). Just in case ideologues jump into the fray, this has nothing to do with capitalism versus socialism or left versus right. It is about planetary justice. It is about making sure people don't go hungry as the environmental 'cake' shrinks. The time is now to ensure that the Earth's resources are equitably shared and not siphoned off by countries who are already overdeveloped and who aspire to be 'great again' at the expense of those less fortunate.

I wonder if there are any doctoral dissertations on the huge profits American multinational companies suck in from the global sales of soft drinks, beef burgers, fried chicken, ice cream, doughnuts, mobile phones, computer software, computer hardware, Hollywood films, etc., not to mention the billions of dollars in advertising revenue from social media. In all these operations, the amount of resources sucked in from poor underdeveloped countries to the US is no less than astronomic. Neo-colonialism is here to stay.

I was toying with the idea of negative growth when the phrase 'degrowth' surfaced in the course of a workshop on sustainable development I attended some years ago. Again it was good to know that I was not out on a limb:

> In the late 1980s, the sustainable development paradigm emerged to provide a framework through which economic growth, social welfare and environmental protection could be harmonized. However, more than 30 years later, we can assert that such harmonization has proved elusive ... The three pillars of sustainability (environment, society and economy) are thus simultaneously threatened by an intertwined crisis.
>
> In an attempt to problematize the sustainable development paradigm, and its recent reincarnation in the concept of a 'green economy', degrowth emerged as a paradigm that emphasizes that there is a contradiction between sustainability and economic growth ... that the pathway towards a sustainable future is to be found in a democratic and redistributive downscaling of the biophysical size of the global economy.[30]

This kind of thinking challenges globalization, as the nation state system propped up by the debt financing of banking is wholly dependent on the growth model. The painful negotiating processes that finally culminated in a replacement to the Kyoto protocol in Paris in 2015 is bound to play itself out again if degrowth ever gets on the international agenda. Vested interests will fight tooth and nail to prevent any such agreement ever seeing the light of day. Who is to say that big players like the United States will not renege on any agreements that may eventually come to fruition, as President Trump did over the Paris Agreement, but by then we might well have already missed the narrow window of opportunity for any meaningful change.

Techno-fixes in the form of geo-engineering are now being proposed as possible solutions to mitigate climate change. One such proposal that will take us to further dizzying heights is to place 'reflectors in space'.[31] However, more grounded scientists are wary about these proposals and those coming up with such bright ideas would be well advised to look up the Earth Charter and the sound advice it gives in following the precautionary principle.[32] This should be required reading for all innovators considering the environmental impact of the heavy human footprint on the planet. It may also put Klaus Schwab, the architect of the Davos Economic Forum, in another frame of mind. He sees the fourth Industrial Revolution led by technology as inevitable and expresses the view that if responsibly handled it 'could catalyse a new cultural renaissance that will enable us to feel part of something much larger than ourselves - a true global civilisation'.[33] At this point we need to consider Pope Francis' view on this matter, which he expresses in his much celebrated *Laudato Si'*:

> It can be said that many problems of today's world stem from the tendency, at times unconscious, to make the method and aims of science and technology an epistemological paradigm which shapes the lives of individuals and the workings of society. The effects of imposing this model on reality as a whole, human and social, are seen in the deterioration of the environment, but this is just one sign of a reductionism which affects every aspect of human and social life. We have to accept that technological products are not neutral, for they create a framework which ends up conditioning lifestyles and shaping social possibilities along the lines dictated by the interests of certain powerful groups. Decisions which may seem purely instrumental are in reality decisions about the kind of society we want to build.[34]

The progress of technology is motivated by profit and not by any altruistic motives, make no bones about it. For example, when the likes of artificial intelligence is marketed, its effects on people and society could have quite profound implications. While the corporations rake in the profits all the resulting social and psychological problems will, as usual, be shouldered by the consumer public. The advent of robotics will intensify concerns for the future of employment and by extension the nature of education. Schwab asserts that, 'we are at the beginning of a revolution that is fundamentally changing the way we live, work, and relate to one another.'[35] This is nothing new, as it has been going on for nigh on 500 years anyway – ways of life redesigned or destroyed. What is new is a concentrated application of these processes which in many ways blows the lid off globalization and exposes it as utterly anti-democratic and largely for the benefit of the powerful. It used to be the case that people and communities decided their futures for themselves. History is also peppered with oppressive regimes, but even then the affair was usually only local. When the nation states arrived they became the agency whereby our destinies were decided, but they also are now locked into the global system. What happened to 'we the people'? To 'equality, fraternity, liberty'? Who is in charge now? The banks? Corporations? The G20 countries? The World Trade Organization? Davos? Ralph Lapp, the nuclear physicist who was part of the Manhattan Project, made this observation:

> No one – not even the most brilliant scientist alive today – really knows where science is taking us ... We are aboard a train which is gathering speed, racing down a track on which there are an unknown number of switches leading to unknown destinations. No single scientist is in the

engine cab and there may be demons at the switch. Most of society is in the caboose looking backward.[36]

My argument is not against science but the use to which it is put.

Capitalism: A debtocracy

In my estimation, globalization began when the Iberian sailors, Columbus and Vasco Da Gama, travelled west and east respectively in the closing decade of the fifteenth century. This process was eventually consolidated when Commander Perry of the United States Navy sailed into Tokyo harbour in 1853 and bullied the Japanese into a trade treaty. This became the first act of globalization, by which western encirclement of the entire globe was completed. The second act of globalization was a programme of cultural encirclement, achieved by the progressive establishment of mission schools across continents as part of the western process of colonization, which still continues to this day. In addition to this, today we have business school 'economic missionaries' busily setting up their branches in the so-called developing world alongside clones of western universities, out of whose halls will emerge acolytes primed to venerate the new technologies as they appear. The third act of globalization occurred when the system of pseudo-money that serves the world today shifted from the private to the public domain with the establishment of the Bank of England in 1694, which enshrined the diabolical fractional reserve system.

These events ensured the continuance of western hegemony and from then onwards we see a process of consolidation, which was initiated paradoxically by the decolonization process. As it received its 'independence', each newly invented country was shackled to the global financial system by a central bank. Further tightening of the grip was achieved by the Bretton Woods agreement in the closing stages of the Second World War in 1944, which was followed by the US decoupling the dollar from gold in 1972. The establishment of the World Trade Organization in 1995 finally sealed the process, the ex-colonial minnows were in the dragnet together with much bigger fish.

I have dealt with the nature of money and banking in previous chapters (see chapter 1, 'A Lethal Cocktail' and chapter 4, 'Acceleration') where I have attempted to show that the essence of money today is that of a virus which can only exist in the nature of debt. All nation states are in debt and here is a sample of figures, in descending order, that represent debt levels as percentage of GDP for 2017 in some of the world's leading economies:[37]

Japan 234.7
Canada 98.8
France 96
United Kingdom 92.2
United States 73.8
Germany 68.2

Then there is also consumer debt. The following is household debt, again in descending order as a percentage of GDP, in 2017 in the same countries:[38]

Canada 99.8
United Kingdom 87.6
United States 78.8
Japan 65.9
France 56.7
Germany 53.4

Norway, which was rated the happiest country in the world by the United Nations in 2017, and Denmark, rated the second, had a household to GDP debt ratio of 98.9 and 123.6 per cent, respectively. 'Household debt is the amount of money all adults in a household owe to any financial institute. Money spent from credit cards, student loans, car loans, home loans, personal loans, and such are cumulatively considered as a household debt.'[39] It would seem that in today's world happiness is anxiety measured by the amount of debt one accumulates to lead consumer lifestyles, and we should by now be able to make the connection between this and global warming: Debt is one of the deepest roots of global warming.

The global banking meltdown, which was created by a debt crisis fabricated by the banks themselves, was at its height in 2008, the same year in which Barack Obama became President of the United States. He thus inherited a crisis of immense proportions, but the officials the erstwhile president appointed to deal with the crisis were the same people who played a not inconsiderable role in creating it in the first place, thus underscoring democracy's dependency on the banks.

Between Adam Smith and Thomas Piketty[40] we've had John Maynard Keynes, who advocated government intervention in the economy, Milton Friedman, who advocated the opposite, and a host of other economists producing their versions of economic theory which in the end has brought the global system to a tipping point. So what's next? Some new thinking is emerging from left-wing intellectuals such as Paul Mason, who is proposing the socialization of the financial system. He argues from a perspective opposite to that of Schwab that, 'Once capitalism can no longer adapt to technological change, postcapitalism becomes necessary'.[41] In his scheme, technology provides incentives for alternative models to prosper, but this ignores the fact that it is capitalism that produces technological change and corporations will still be in control. Pope Francis observes that:

> ... there is urgent need for politics and economics to enter into a frank dialogue ... Saving banks at any cost, making the public pay the price, foregoing a firm commitment to reviewing and reforming the entire system, only reaffirms the absolute power of a financial system, a power which has no future and will only give rise to new crises after a slow, costly

and only apparent recovery. The financial crisis of 2007–08 provided an opportunity to develop a new economy, more attentive to ethical principles, and new ways of regulating speculative financial practices and virtual wealth. But the response to the crisis did not include rethinking the outdated criteria which continue to rule the world.[42]

Perhaps this is as far as the good Pope would go using the moderated language that is appropriate to his station. But there are more trenchant Christian voices. My wife Saba and I spent a very interesting time in the mid-1990s working with the Christian Council for Monetary Justice (CCMJ), lobbying members of Parliament in London on these issues. The CCMJ is scathing about the 'unquestioning acceptance of the economic status quo [and the] fraudulent, irreligious banking, money/credit debt system. CCMJ encourages people to explore for themselves the mystery of poverty in the world of actual and potential abundance'.[43] They advocate a system which transfers the power of money creation to the state, the theoretical approach to which is argued by the Canadian economist John Tomlinson.[44]

As I write this (January 2018), the *Guardian* newspaper carries this headline on its front page: '"Doomsday Clock" ticked forward 30 seconds to 2 minutes to midnight'. It was reporting the release of the Bulletin of the Atomic Scientists, 'in a reflection of how the scientists view the dangers facing the world'.[45] This was yet another warning by scientists in just a matter of weeks that there is something seriously awry with the way we are running the world. Nuclear Armageddon is their first concern, given the unpredictable leadership both in the east and the west, and the other issue it singles out is climate change: 'avoiding catastrophic temperature increases in the long run requires urgent attention now ... and so far, the global response has fallen far short of meeting this challenge.'[46]

On this very same day, President Trump of the United States made a major speech at the Davos Economic Forum inviting world leaders to do business with America. As the champion of economic growth this by all accounts went down well with the business community, but there was no reference to climate change, and the fact was that he had led his country to renege on the climate treaty concluded in Paris in 2015. This was also the day when the New York stock exchange registered record rises in share values (a downturn has occurred since then) and the UK reported higher economic growth, with some self-satisfaction.

Needless to say, all Muslim countries – regardless of their Islamic credentials – are also locked into this global system. They have no choice in the matter and in many ways they also want most things the system has to offer, even nuclear reactors. If there is an institution that claims to provide an alternative financial route for Muslims it is the Islamic Bank, which by its very definition is an oxymoron. It is joined at the hip to the global mainstream banking system and owes its existence to theological compromises conjured up by certain scholars in their eagerness to be modern. As the Qur'an asserts: *Those who take ribā (usury/interest) will rise up on the Day of Resurrection like someone tormented by Satan's touch* (2: 275).

Where do we go from here? Primarily, it is obvious that we need to abandon our addiction to economic growth. The idea of 'equity' needs to replace our outdated economic model defined in terms of prosperity, progress and development. New thinking demands that we address our minds to 'over development' and consider the idea of degrowth, briefly discussed above. There will no doubt be resistance to these ideas from the political establishment, banks, economists and their fellow travellers. Developed countries and those playing catch up, led by China and India, whose growth rates are alarming, will no doubt join the fray. The aspirations they have for their populations are laudable but the methods they employ can only be counterproductive, as we are already beginning to see in their smog-covered cities. If they deliver half the living standards they have in mind, in comparison to the developed countries, the burden on both renewable and non-renewable resources would be such that we are assured of a collapse in Earth systems. The aspirations of the people of developing nations for a better life can only be met, given the current state of the planet, if the problem of over development in the richer countries can be resolved. A logical and reasonable position to take that has in my view been scrupulously avoided up to the present time. This could be regarded as political dynamite, but given that global resources are finite would it not be sensible, fair and prudent even at this late stage to define an optimum state of wellbeing that will apply to the entire planet, beyond which a nation state would be considered to be over developed? Of course it would, but for this idea to be accepted even as a starting point for discussion there needs to be a radical shift in attitudes, particularly amongst the developed nations. I am probably asking too much, as there will always be those who want to grab more than their fair share of the cake, as evidenced by the fact that the biggest contributors to global warming are the so called developed nations. A study by Oxfam, the International charity, has found that the 'poorest half of the global population – around 3.5 billion people – are responsible for only around 10% of total global emissions attributed to individual consumption'.[47] If it is global equity we are looking for 'the pathway towards a sustainable future is to be found in a democratic and redistributive downscaling of the biophysical size of the global economy'.[48]

Capital, which is now represented by algorithms generated by the banks, feeds growth and sustainable growth is another oxymoron. It is the back door to overdevelopment, which is one of the root of climate change, and economists are now coming up with ideas that promote carbon-free growth. This studiously avoids the implications this approach will have on scarce and non-renewable resources, the resource demands of globalized economic activity, and the further degradation of the planet. But there is a way out, if only it is grasped with the required good will and openness, in spite of the approaching bleak scenario that has been painted for us by scientists and academics. The international community has successfully negotiated a Framework Convention on Climate Change (UNFCCC), so why not a United Nations Framework Convention on

Equity, where fairness and eco-justice replace greed, selfishness and narrow nationalism? I can see that this is going to make huge demands on the altruistic side of our natures, but is this too much to ask on behalf of the children who will inherit our Earth? This is possibly an area where FBOs can come into their own, if they will only stand up for it. The current economic model needs also to be replaced with another model, whereby the local receives greater emphasis than the global. In this scenario, the demeaning term 'economic migrant' will be lost to history, as all people see the possibility and value of reviving their own local environments and economies. The banking function of money creation should be transferred to the state, and this activity progressively reduced and finally eliminated. If we can fly to the moon and back I am sure we can work this one out. But again, vested interest and narrow nationalism will attempt to stand in the way. In this regard, The New Economics Foundation (NEF), the UK based independent think-tank, should be congratulated for giving a push start to this discussion. It observes:

> ... nothing short of a *Great Transition* to a new economy is necessary and desirable, and also possible. Business as usual has failed. Yet prime ministers, finance ministers and governors of central Banks are still running around ... trying to allay fears and convince us that this is not the case.[49]

But can we do this before the cashless society emerges and the Blockchain online system[50] and Bitcoins[51] engulf us in their vortex? If the bath is overflowing, the sensible thing to do is to turn off the tap. The overflow of 'virus' money needs to be controlled and eventually eliminated. Not only is this possible, but examples of creative economic activities already exist. With an eighty-year history, the Swiss WIR Bank is one such independent alternative currency system.[52] There is also an encouraging account of how a community on the island of Sardinia lifted itself out of a dire economic situation by creating their own exchange system.[53] A group of Muslims networked internationally have minted gold and silver coins based on the template defined by the first Muslim community in Madinah during the time of Prophet Muhammed.[54]

The problem is systemic and a paradigm shift involving lifestyle change is now urgently called for if we are to avoid planetary collapse. A blocking of change by powerful vested interests is the main barrier to this shift in the level of consciousness. But it is also an opportunity for faith communities to regain the initiative they have lost. This will happen by working collectively and joining forces wherever required, and there is also much that they can do to infuse a sense of balance and hold in check the extreme and adverse manifestations of modernity within their own spheres of influence.

Hope Springs Eternal[55]

The turning

We are at a fork in the road.

We can ignore the turn and continue in the same direction as always, with faith in the progress that has brought those creature comforts that are now commonplace, especially to those of us who live in the so-called developed world. Our hopes run high and we have convinced ourselves that the civilization we have created is never going to end. We have contrived a set of values that we can summon to justify the direction we are taking and convince ourselves that it is all going to get better, and better still. The pleasure principle and consumerism drive our aspirations to the extent of making plans to exploit the solar system for minerals.

While we aspire to look for resources in space we continue to destroy life in our own back yard. We ignore the mismatch between our resource-hungry lifestyles and the state of planet Earth, on the one hand, and our ambitious aspirations for the future on the other. It is the height of irresponsibility to continue with the language of growth, whichever quarter it comes from, and concoct pseudo-economic theories to justify this position. Perhaps it is time to re-examine the idea of entropy, which emerges from the second law of thermodynamics, 'that everything in the universe eventually moves from order to disorder'.[56] Cultural or social entropy suggests that society too has an end date. Civilizations reach their peak then progressively become more frenetic, violent and collapse. It seems to me that we are going through this process, now gripped by a sense of anomie. Our civilization, which has ostensibly been built on the scientific paradigm, ignores its own science and the warnings of its own scientists.

Or we take the turning at the fork, and having bruised and battered the natural world for so long once again snuggle up to her warm embrace as a child does to its mother, declare our peace with her and restore the equilibrium between us. We are inescapably interwoven into her fabric, as the fresh air she provides fills our lungs, the fresh water she provides infuses our bodies and quenches our thirst, the trace elements in our bodies: iron, calcium, magnesium, phosphorous, potassium, keep our beings in balance. She nurtures us by her abundance but she is not a resource for our profligacy. We are now numerous and there are more than 7.6 billion of us and growing. In a saying attributed to Prophet Muhammad, he urges people to continue planting trees even if they think it is the last day on Earth. This manifests a sense of optimism and hope, and also one of determination to remain positive to the end if need be.

Working together

We find ourselves in a situation that is unprecedented. In spite of their varying world views, faith communities have, after all, been chasing the same thing:

Prosperity. In addition to health, our prayers almost invariably contain an appeal for wealth. Along comes the Pied Piper of Hamelin, who shows us a shortcut to the banks, and we now know where that has taken us. The point is we are collectively responsible, and having been involved for nigh on forty years in the interfaith environment movement I have to say that we have no other option but to work together in finding answers to what we have wrought. We need to restore the divine and what is sacred to their rightful positions in the scheme of our lives. Faith Based Organizations have emerged as a vocal lobbying group, particularly in international fora associated with climate change. They have earned the recognition of the United Nations and although they still lack coherence they are a growing movement of global proportions like no other.

I alluded in the previous chapter to the Ohito Declaration for Religion Land and Conservation. It contained ten common environmental principles and ten areas of action where the participating traditions found common ground. I reproduce here the set of common principles we arrived at:

1) Religious beliefs and traditions call us to care for the Earth.
2) For people of faith maintaining and sustaining environmental life systems is a moral responsibility.
3) Nature should be treated with respect and compassion, thus forming a basis for our sense of responsibility for conserving plants, animals, land, water, air and energy.
4) Environmental understanding is enhanced when people learn from the example of prophets, teachers and of nature itself.
5) Markets and trade arrangements should reflect the physical and spiritual needs of people and their communities to ensure health, justice and harmony. Justice and equity principles of faith traditions should be used for maintaining and sustaining environmental life systems.
6) People of faith should give more emphasis to a higher quality of life in preference to a higher standard of living, recognizing that greed and avarice are the root causes of environmental degradation and human debasement.
7) All faiths should fully recognize and promote the role of women in environmental sustainability.
8) People of faith should be fully involved in the conservation and development process. Development of the environment must take better account of its effects on the community and its religious beliefs.
9) Faith communities should endorse multilateral consultation in a form that recognizes the value of local/indigenous wisdom and current scientific information.
10) In the context of faith perspective, emphasis should be given not only to the globalization of human endeavours but also to participatory community action.

This is only a start and there is no shortage of things we could be doing together. The Islamic Foundation for Ecology and Environmental Sciences has been in the forefront of developing an Islamic environmentalism since the 1990s. Amongst numerous other projects, it put an end to dynamite fishing in Zanzibar in 1998; carried out workshops based on Islamic environmental teachings from West Africa to Indonesia; it has been at the core of the production of the Islamic Global Declaration on Climate Change (see appendix) working with Islamic Relief, Climate Action Network International and Green Faith and it is now part of the coalition of FBOs dealing with this issue; it was engaged in 2017 to support a research study related to conserving coral reefs in the Western Indian Ocean, working with Plymouth Marine Lab and Exeter University.[57] It is also now part of a coalition of FBOs set up by the Norwegian government and the United Nations Development Programme to conserve tropical rainforests. I urge Muslims, both young and old, to be part of this movement and work at whatever level that is appropriate for them. There are organizations both local and national that could do with being supported; fresh ideas, fresh blood, fresh enthusiasm is what is needed and there will assuredly be a welcome at the door.

There are two ideas which are gaining momentum I would like to flag at this point. The first is the 'transition town' movement, where concerned people are working together looking at adaptation in post-fossil fuel societies:

One of the defining aims of a typical Transition Town (or Transition Initiative, as they are also known) is to relocalize and decarbonize the economy, in order to become less dependent on the globalized, oil-dependent economy. This involves coming together as a community with the ambitious, long-term goal of using mostly local resources to meet local needs. For example, rather than relying on industrially produced food that is imported from all around the world, Transition Towns try to maximize local, organic food production and exchange. Rather than relying on fossil fuel energy, Transition Towns take steps to radically reduce energy consumption while moving to renewable sources. Rather than mindlessly embracing consumerism, participants in Transition Towns are stepping out of the rat race, reimagining 'the good life,' and discovering that community engagement provides great wealth to those brave enough to get involved.[58]

The second is the idea of the 'circular economy', an alternative to a traditional linear economy (make, use, dispose) in which we keep resources in use for as long as possible, extract the maximum value from them whilst in use, then recover and regenerate products and materials at the end of each service life.'[59]

These ideas are very much in keeping with what Islam teaches and what I have been advocating for nigh on forty years.

The Islamic contribution

I have attempted to show in the previous chapter ('Islam and the Natural World') that what we have come to describe today as environmentalism is deeply embedded in the matrix of Islam. It is at its simplest level about good manners. It is about personal behaviour and how it manifests itself in our relationships with others. It is about being well behaved in our relationship with the natural world and other sentient beings. The exemplar is Prophet Muhammad, and it grew from the foundations he established into a range of rules and institutions that manifested an expression of life in a way that is truly holistic. As the prophetic model is based on the Qur'an it could be distilled into three categories, bearing in mind that public good must be the ultimate objective. They are to do what is right, forbid what is wrong and act with moderation at all times: *Let there be a community among you that calls for what is good, urges what is right and forbids what is wrong; those are the ones who have success* (3: 104).

The body of the Shariah allows us to deduce three general principles as it pertains to the natural world:

- The elements that compose the natural world are common property, bearing in mind that the ultimate owner of the Earth is the Creator.
- The right to benefit from natural resources is a right held in common.
- There shall be no damage or infliction of damage to it bearing in mind future users.

What is now emerging as 'Islamic environmentalism' (a tautology) originated from this foundational code and Muslim legalists have over the centuries worked out both principles and structures to give this expression. They concern individual rights, obligations and responsibilities individuals owe to the community, accountability, benefits accruing to users from renewable and non-renewable resources held in common, and penalties for improper use of natural resources. I have given concrete examples in the previous chapter of how this could work, although I see it only as a beginning. My intentions have been to start a process of dialogue and to invite scholars with greater understanding of the Qur'an and the traditions to participate in extending and improving this knowledge base, as there is much to be drawn from the sources.

But I cannot repeat often enough that there is an inherent urgency in what we face given the predictions by scientists of global systems collapse and the looming climate crisis. What Muslims, who form over one-fifth of the world's population, can offer the rest to mitigate the collapse and how soon we do it will have a bearing on how the human race will survive in a changed world. Equally, how the rest relate to planet Earth will have a bearing on Muslims, and the times call for a sensitivity to these common challenges in a shared space. I have laid out a template (see chapter 5, 'Political Economy') where we could lead reasonably satisfactory lifestyles that meet our needs based on the prophetic tradition, where caring and sharing takes precedence over selfishness, personal aggrandizement and greed. I summarize them below:

- Live moderate simple lives.
- Do not hoard wealth for its own sake.
- Share your wealth. Pay the compulsory zakat – 2½ per cent of your savings – as charity. Give your surplus wealth to the needy and good causes.
- Caring for the environment is a sacred undertaking.
- Broadly-speaking, money is any non-perishable commodity that can be used as a medium of exchange. Usury/interest and fake money is forbidden.
- Markets must be free and open.
- Fair trade is encouraged and is built into contractual obligations.
- Wages must be paid immediately when they are due.
- Conduct affairs by mutual consultation.
- Give precedence to the rule of law.

I believe that the time for empire, hegemony and the grand gesture is at an end (in spite of what is happening in the United States today) and the case for simplicity is unassailable. Small, inevitably, is beautiful, as was the first community of Prophet Muhammad in Madinah. That was a social pattern Muslims can emulate, working in harmony with the heartbeat of the natural world. It can be applied at the level of the nation state and also in small communities wherever Muslims live. There are good examples to be set.

I conclude as I began with the Qur'an:

He (Allah) set up the balance
So that you may not exceed the balance,
weigh with justice and do not fall short in the balance.
He (Allah) has spread out the earth for living creatures
With its fruit and date-palms with sheathed clusters,
also grain on leafy stems and fragrant herbs.
Then which of the favours of your Lord will you deny? (55: 7–13)

Endnotes

1. The Millennium Alliance for Humanity and the Biosphere (MAHB) is run by a group of intellectuals headed by Paul Ehrlich, who is Professor of Population Studies and President, Center for Conservation Biology, Stanford University, USA.
2. Jeremy Lent, 'What Will It Really Take to Avoid Collapse?' (MAHB, 2 January 2018). Available at: http://mahb.stanford.edu/blog/avoid-collapse/ (accessed on 9 May 2018).

3. Noel Castree, 'Has Planet Earth Entered New "Anthropocene" Epoch?' (Live Science, 30 August 2016). Available at: http://www.livescience.com/55942-has-planet-earth-entered-new-anthropocene-epoch.html (accessed on 16 July 2018).

4. Op cit.

5. 'Abu Dawud: Book 016, Hadith Number 3325' (Hadith Collection, undated). Available at: http://www.hadithcollection.com/abudawud/248-Abu%20Dawud%20Book%2016.%20Commercial%20Transactions/17473-abu-dawud-book-016-hadith-number-3325.html (accessed on 29 May 2018).

6. Castree, 'Has Planet Earth Entered New "Anthropocene" Epoch?'

7. William J Ripple, Christopher Wolf, Thomas M Newsome, et al., 'World Scientists' Warning to Humanity: A Second Notice' *BioScience*, 67/12 (1 December 2017), pp. 1026–1028. Available at: https://academic.oup.com/bioscience/article/67/12/1026/4605229 (accessed on 9 May 2018).

8. Op cit.

9. Op cit.

10. Op cit.

11. Op cit.

12. Jeremy Lent, *The Patterning Instinct: A Cultural History of Humanity's Search for Meaning* (New York: Prometheus Books, 2017). See also Jeremy Lent, 'The Patterning Instinct: A Cultural History of Humanity's Search for Meaning' (MAHB, 23 May 2017). Available at: https://mahb.stanford.edu/recentnews/patterning-instinct-cultural-history-humanitys-search-meaning/ (accessed on 9 May 2018).

13. Lent, 'What Will It Really Take to Avoid Collapse?'

14. Summary for Policymakers of IPCC Special Report on Global Warming of 1.5oC approved by governments (IPCC PRESS RELEASE 8 October 2018). Available at https://ipcc.ch/pdf/session48/pr_181008_P48_spm_en.pdf. (Accessed 9 October 2018).

15. 'SIPRI Yearbook 2017: Armaments, Disarmament and International Security' (Stockholm: SIPRI, 2017). Available at: https://www.sipri.org/sites/default/files/2017-09/yb17-summary-eng.pdf (accessed on 9 May 2018).

16. Andrew Roth, 'Putin threatens US arms race with new missiles declaration' (*The Guardian*, 1 March 2018). Available at: https://www.theguardian.com/world/2018/mar/01/vladimir-putin-threatens-arms-race-with-new-missiles-announcement (accessed on 9 May 2018).

17. Larry Elliott, Richard Partington, and Edward Helmore, 'US on brink of trade war with EU, Canada and Mexico as tit-for-tat tariffs begin' (*The Guardian*, 31 May 2018). Available at: https://www.theguardian.com/business/2018/may/31/us-fires-opening-salvo-in-trade-war-with-eu-canada-and-mexico (accessed on 14 June 2018); Robert Delaney, 'Donald Trump vows to pursue aggressive trade action in 'coming weeks' against China' (*South China Morning Post*, 14 June 2018). Available at: http://www.scmp.com/news/china/diplomacy-defence/article/2150689/donald-trump-vows-pursue-aggressive-trade-action-coming (accessed on 14 June 2018).

18. Press Release, 'Richest 1 percent bagged 82 percent of wealth created last year – poorest half of humanity got nothing' (OXFAM International, 22 January 2018). Available at: https://www.oxfam.org/en/pressroom/pressreleases/2018-01-22/richest-1-percent-bagged-82-percent-wealth-created-last-year (accessed on 10 May 2018).

19. Lara Nassar and Juliet Dryden (ed.), *The Arab Environmental Governance Charter* (West Asia–North Africa Institute, December 2016). Available at: http://wanainstitute.org/en/publication/arab-environmental-governance-charter (accessed on 27 July 2018).

20. Jeff Farrell, 'Pakistani province plants one billion trees to help slow down effects of global warming' (*Independent*, 14 August 2017). Available at: http://www.independent.co.uk/news/world/asia/pakistan-plant-billion-trees-global-warming-effects-climate-change-imran-khan-khyber-pakhtunkhaw-a7892176.html (accessed on 10 May 2018).

21. 'South Africa: Cape Town slashes water use amid drought' (BBC News, 18 January 2018). Available at: http://www.bbc.co.uk/news/world-africa-42731084 (accessed on 10 May 2018).

22. John Vidal, 'Who should feed the world: real people or faceless multinationals?' (*The Guardian*, 5 June 2018). Available at: https://www.theguardian.com/commentisfree/2018/jun/05/feed-the-world-real-people-faceless-multinationals-monsanto-bayer (accessed on 14 June 2018).

23. Ponting, *A New Green History of the World*, p. 125.

24. Eric Johnson, 'Where Would We Be Without Banks?' (Credit Suisse, 22 July 2015). Available at: https://www.credit-suisse.com/corporate/en/articles/news-and-expertise/where-would-we-be-without-banks-201507.html (accessed on 17 July 2018).

25. 'World Footprint' (Global Footprint Network, undated). Available at: https://www.footprintnetwork.org/our-work/ecological-footprint/#worldfootprint (accessed on 10 May 2018).

26. Simms and Johnson, *Growth isn't possible*, p. 5.

27. *World Economic Outlook, October 2017, Seeking Sustainable Growth: Short-Term Recovery, Long-Term Challenges, October 2017*. Available at HTML https://www.imf.org/en/Publications/WEO/Issues/2017/09/19/world-economic-outlook-october-2017 (accessed on 17 July 2018)

28. 'Overview' in *Global Economic Prospects: Broad-Based Upturn, but for How Long?* (The World Bank, undated). Available at: http://www.worldbank.org/en/publication/global-economic-prospects (accessed on 10 May 2018). The full report may be obtained from the same link.

29. 'United Nations Sustainable Development Goals: 17 goals to transform our world' (UN, undated). Available at: http://www.un.org/sustainabledevelopment/sustainable-development-goals/ (accessed on 10 May 2018).

30. Viviana Asara, Iago Otero, Federico Demaria and Esteve Corbera, 'Socially sustainable degrowth as a social–ecological transformation: repoliticising sustainability', *Sustainability Science*, 10/3 (July 2015), pp. 375–384. Available at: https://link.springer.com/article/10.1007/s11625-015-0321-9 (accessed on 10 May 2018).

31. 'Geoengineering to Combat Global Warming: Why is this issue important?' (UNEP, May 2011). Available at: https://na.unep.net/geas/getUNEPPageWithArticleIDScript.php?article_id=52 (accessed on 10 May 2018).

32. 'The Earth Charter' (Earth Charter Initiative, undated). Available at: http://earthcharter.org/discover/the-earth-charter/ (accessed on 10 May 2018).

33. Klaus Schwab, *The Fourth Industrial Revolution*. (Cologny/Geneva Switzerland, 2016) p. 114.

34. Libreria Editrice Vaticana, 'Encyclical Letter *Laudato Si*", #107.

35. Schwab, op cit. p. 1.

36. Quoted in Alvin Toffler, *Future Shock* (New York: Pan Books, 1971), p. 390. Ralph Lapp was a nuclear physicist who was part of the Manhattan Project that produced the first atomic bomb.

37. Tejvan Pettinger, 'List of National Debt by Country' (Economics.help, 1 June 2017). Available at: https://www.economicshelp.org/blog/774/economics/list-of-national-debt-by-country/ (accessed on 10 May 2018).

38. Wolf Richter, 'Here are the countries with the biggest debt slaves' (Business Insider, 23 January 2017). Available at: http://www.businessinsider.com/these-are-the-countries-with-the-biggest-debt-slaves-2017-1?IR=T (accessed on 10 May 2018).

39. Jack Massey, 'Top 10 Countries Having The Highest Household Debt to GDP Ratio in 2017' (More Buzzing, 3 February 2017). Available at: http://www.morebuzzing.com/top-10-countries-having-the-highest-household-debt-to-gdp-ratio-in-2017/ (accessed on 17 July 2018).

40. Thomas Piketty and Arthur Goldhammer (trans.), *Capital in the Twenty-First Century* (Cambridge, MA: Harvard University Press, 2014).

41. Paul Mason, *Postcapitalism: A Guide to Our Future* (London: Allen Lane, 2015), p. xiii.

42. Libreria Editrice Vaticana, 'Encyclical Letter *Laudato Si*", #189.

43. 'Organisation' (Christian Council for Monetary Justice, undated). Available at: http://ccmj.org/wiki/view.php?page=Organisation (accessed on 10 May 2018).

44. John Tomlinson, *Honest Money: A Challenge to Banking* (Deddington, Oxon.: Helix, 1993).

45. Julian Borger, '"Doomsday Clock" ticked forward 30 seconds to 2 minutes to midnight' (*The Guardian*, 26 January 2018). Available at: https://www.theguardian.com/world/2018/jan/25/doomsday-clock-ticked-forward-trump-nuclear-weapons-climate-change (accessed on 10 May 2018).

46. John Mecklin (ed.), 'It is now two minutes to midnight: 2018 Doomsday Clock Statement' (Bulletin of the Atomic Scientists, 25 January 2018). Available at: https://thebulletin.org/2018-doomsday-clock-statement (accessed on 10 May 2018).

47. Oxfam media briefing, 'Extreme Carbon Inequality' (Oxfam, 2 December 2015). Available at: https://www.oxfam.org/sites/www.oxfam.org/files/file_attachments/mb-extreme-carbon-inequality-021215-en.pdf (accessed on 14 June 2018).

48. Asara, et al., op cit.

49. Stephen Spratt and Mary Murphy, *The Great Transition: A tale of how it turned out right* (London: New Economics Foundation, 2009), p. 3.

50. Simply described Blockchains are ostensibly incorruptible electronic ledgers about to take the world by storm.

51. Bitcoin is a digital currency that works as a worldwide payment system without banks.

52. Anthony Migchels, 'The Swiss WIR, or: How to Defeat the Money Power' (Real Currencies, 19 April 2012). Available at: https://realcurrencies.wordpress.com/2012/04/19/the-swiss-wir-or-how-to-defeat-the-money-power/ (accessed on 10 May 2018).

53. Edward Posnett, 'The Sardex Factor' (*Financial Times*, 18 September 2015). Available at: https://www.ft.com/content/cf875d9a-5be6-11e5-a28b-50226830d644 (accessed on 10 May 2018).

54. Umar Ibrahim Vadilo, *The Return of the Gold Dinar: A study of money in Islamic law and the architecture of the gold economy* (Kuala Lumpur: Madinah Press, 2004).

55. With thanks to Alexander Pope, from 'An Essay on Man' (1733).

56. 'Entropy' (Merriam Webster, undated). Available at: https://www.merriam-webster.com/dictionary/entropy (accessed January 2018).

57. 'Coral Communities: Building Socio-Ecological Resilience to Coral Reef Degradation in the Islands of the Western Indian Ocean' (Plymouth Marine Laboratory, undated). Available at: http://www.pml.ac.uk/Research/Projects/Coral_Communities (accessed on 10 May 2018).

58. 'What is a Transition Town?' (Permaculture Research Institute, undated). Available at: https://permaculturenews.org/2012/07/21/what-is-a-transition-town/ (accessed on 10 May 2018).

59. 'WRAP and the circular economy' (Waste and Resources Action Programme, undated). Available at: http://www.wrap.org.uk/about-us/about/wrap-and-circular-economy (accessed on 10 May 2018).

APPENDIX

ISLAMIC DECLARATION ON GLOBAL CLIMATE CHANGE

In the name of Allah, Most Merciful, Most Compassionate

PREAMBLE

1.1 God – Whom we know as Allah – has created the universe in all its diversity, richness and vitality: the stars, the sun and moon, the earth and all its communities of living beings. All these reflect and manifest the boundless glory and mercy of their Creator. All created beings by nature serve and glorify their Maker, all bow to their Lord's will. We human beings are created to serve the Lord of all beings, to work the greatest good we can for all the species, individuals, and generations of God's creatures.

1.2 Our planet has existed for billions of years and climate change in itself is not new. The earth's climate has gone through phases wet and dry, cold and warm, in response to many natural factors. Most of these changes have been gradual, such that the forms and communities of life have adjusted accordingly. There have been catastrophic climate changes that brought about mass extinctions, but over time, life adjusted even to these impacts, flowering anew in the emergence of balanced ecosystems such as those we treasure today. Climate change in the past was also instrumental in laying down immense stores of fossil fuels from which we derive benefits today. Ironically, our unwise and short-sighted use of these resources is now resulting in the destruction of the very conditions that have made our life on Earth possible.

1.3   The pace of Global climate change today is of a different order of magnitude from the gradual changes that previously occurred throughout the most recent era, the Cenozoic. Moreover, it is human-induced: we have now become a force dominating nature. The epoch in which we live has increasingly been described in geological terms as the Anthropocene, or 'Age of Humans'. Our species, though selected to be a caretaker or steward (*khalifah*) on the earth, has been the cause of such corruption and devastation on it that we are in danger ending life as we know it on our planet. This current rate of climate change cannot be sustained, and the earth's fine equilibrium (*mīzān*) may soon be lost. As we humans are woven into the fabric of the natural world, its gifts are for us to savour. But the same fossil fuels that helped us achieve most of the prosperity we see today are the main cause of climate change. Excessive pollution from fossil fuels threatens to destroy the gifts bestowed on us by God – gifts such as a functioning climate, healthy air to breathe, regular seasons, and living oceans. But our attitude to these gifts has been short-sighted, and we have abused them. What will future generations say of us, who leave them a degraded planet as our legacy? How will we face our Lord and Creator?

1.4   We note that the Millennium Ecosystem Assessment (UNEP, 2005), backed by over 1,300 scientists from 95 countries, found that 'overall, people have made greater changes to ecosystems in the last half of the twentieth century than at any time in human history ... these changes have enhanced human well-being, but have been accompanied by ever increasing degradation (of our environment).'

'Human activity is putting such a strain on the natural functions of the earth that the ability of the planet's ecosystems to sustain future generations can no longer be taken for granted.'

1.5   Nearly ten years later, and in spite of the numerous conferences that have taken place to try to agree on a successor to the Kyoto Protocol, the overall state of the earth has steadily deteriorated. A study by the Intergovernmental Panel on Climate Change (IPCC) comprising representatives from over 100 nations, published in March 2014, gave five reasons for concern. In summary, they are:

• Ecosystems and human cultures are already at risk from climate change;
• Risks resulting from climate change caused by extreme events such as heat waves, extreme precipitation and coastal flooding are on the rise;

- These risks are unevenly distributed, and are generally greater for the poor and disadvantaged communities of every country, at all levels of development;
- Foreseeable impacts will affect adversely the earth's biodiversity, the goods and services provided by our ecosystems, and our overall global economy;
- The earth's core physical systems themselves are at risk of abrupt and irreversible changes.

We are driven to conclude from these warnings that there are serious flaws in the way we have used natural resources – the sources of life on Earth. An urgent and radical reappraisal is called for. Humankind cannot afford the slow progress we have seen in all the COP (Conference of Parties – climate change negotiations) processes since the Millennium Ecosystem Assessment was published in 2005, or the present deadlock.

1.6   In the brief period since the Industrial Revolution, humans have consumed much of the non-renewable resources which have taken 250 million years to produce in the earth – all in the name of economic development and human progress. We note with alarm the combined impacts of rising per capita consumption together with the rising human population. We also note with alarm the multi-national scramble now taking place for more fossil fuel deposits under the dissolving ice caps in the arctic regions. We are accelerating our own destruction through these processes.

1.7   Leading climate scientists now believe that a rise of two degrees centigrade in global temperature, which is considered to be the 'tipping point', is now very unlikely to be avoided if we continue with business-as-usual; other leading climate scientists consider 1.5 degrees centigrade to be a more likely 'tipping point'. This is the point considered to be the threshold for catastrophic climate change, which will expose yet more millions of people and countless other creatures to drought, hunger and flooding. The brunt of this will continue to be borne by the poor, as the earth experiences a drastic increase in levels of carbon in the atmosphere brought on in the period since the onset of the Industrial Revolution.

1.8   It is alarming that in spite of all the warnings and predictions, the successor to the Kyoto Protocol which should have been in place by 2012, has been delayed. It is essential that all countries, especially the more developed nations, increase their efforts and adopt the pro-active approach needed to halt and hopefully eventually reverse the damage being wrought.

WE AFFIRM

2.1   We affirm that Allah is the Lord and Sustainer (*Rabb*) of all beings:

الْحَمْدُ لِلَّـهِ رَبّ الْعَالَمِينَ

Praise be to Allah, Lord and Sustainer of all beings.

(Qur'an 1: 1)

He is the One Creator – He is *Al-Khāliq*:

هُوَ اللَّهُ الْخَالِقُ الْبَارِئُ الْمُصَوِّرُ

He is Allah – the Creator, the Maker, the Giver of Form.

(Qur'an 59: 24)

الَّذِي أَحْسَنَ كُلَّ شَيْءٍ خَلَقَهُ

He Who has perfected every thing He has created.

(Qur'an 32: 7)

Nothing that He creates is without value: each thing is created *bi 'l-haqq*, in truth and for right.

وَمَا خَلَقْنَا السَّمَاوَاتِ وَالْأَرْضَ وَمَا بَيْنَهُمَا لَاعِبِينَ مَا خَلَقْنَاهُمَا إِلَّا بِالْحَقّ

And We did not create the heavens and earth and all that is between them in jest.

We have not created them but in truth.

(Qur'an 44: 38–39)

2.2   We affirm that He encompasses all of His creation – He is *al-Muhit*.

وَلِلَّهِ مَا فِي السَّمَاوَاتِ وَمَا فِي الْأَرْضِ وَكَانَ اللَّهُ بِكُلِّ شَيْءٍ مُحِيطًا

All that is in the heavens and the earth belongs to Allah.

Allah encompasses all things.

(Qur'an 4: 126)

2.3   We affirm that:

- God created the earth in perfect equilibrium (*mīzān*);
- By His immense mercy we have been given fertile land, fresh air, clean water and all the good things on Earth that make our lives here viable and delightful;
- The earth functions in natural seasonal rhythms and cycles: a climate in which living beings – including humans – thrive;
- The present climate change catastrophe is a result of the human disruption of this balance.

وَالسَّمَاء رَفَعَهَا وَوَضَعَ الْمِيزَانَ

أَلاَّ تَطْغَوْا فِي الْمِيزَانِ

وَأَقِيمُوا الْوَزْنَ بِالْقِسْطِ وَلا تُخْسِرُوا الْمِيزَانَ

وَالأَرْضَ وَضَعَهَا لِلأَنَامِ

He raised the heaven and established the balance

so that you would not transgress the balance.

Give just weight – do not skimp in the balance.

He laid out the earth for all living creatures.

(Qur'an 55: 7–10)

2.4   We affirm the natural state (*fitrah*) of God's creation:

فَأَقِمْ وَجْهَكَ لِلدِّينِ حَنِيفًا فِطْرَةَ اللَّهِ الَّتِي فَطَرَ النَّاسَ عَلَيْهَا

لا تَبْدِيلَ لِخَلْقِ اللَّهِ ذَلِكَ الدِّينُ الْقَيِّمُ وَلَكِنَّ أَكْثَرَ النَّاسِ لا يَعْلَمُونَ

So set your face firmly to the faith in pure devotion,

the natural pattern on which Allah made humankind.

There shall be no changing Allah's creation.

That is the true Way,

but most people do not know.

(Qur'an 30: 30)

2.5   We recognize the corruption (*fasad*) that humans have caused on Earth in our relentless pursuit of economic growth and consumption. Its consequences have been:

- Global climate change, which is our present concern, in addition to:
- Contamination and befoulment of the atmosphere, land, inland water systems, and seas;
- Soil erosion, deforestation and desertification;
- Destruction, degradation, and fragmentation of the habitats of the earth's communities of life, with devastation of some of the most biologically diverse and productive ecosystems such as rainforests, freshwater wetlands, and coral reefs;
- Impairment of ecosystem benefits and services;
- Introduction of invasive alien species and genetically modified organisms;
- Damage to human health, including a host of modern-day diseases.

ظَهَرَ الْفَسَادُ فِي الْبَرِّ وَالْبَحْرِ بِمَا كَسَبَتْ أَيْدِي النَّاسِ لِيُذِيقَهُم بَعْضَ الَّذِي عَمِلُوا لَعَلَّهُمْ يَرْجِعُونَ

Corruption has appeared on land and sea

by what people's own hands have wrought,

that He may let them taste some consequences of their deeds,

so that they may turn back.

(Qur'an 30: 41)

2.6   We recognize that we are but a minuscule part of the divine order, yet within that order we are exceptionally powerful beings, and have the responsibility to establish good and avert evil in every way we can. We also recognize that:

- We are but one of the multitude of living beings with whom we share the earth;
- We have no right to abuse the creation or impair it;
- Intelligence and conscience should lead us, as our faith commands, to treat all things with care and awe (*taqwa*) of their Creator, compassion (*rahmah*) and utmost good (*ihsān*).

وَمَا مِن دَآبَّةٍ فِي الأَرْضِ وَلاَ طَائِرٍ يَطِيرُ بِجَنَاحَيْهِ إلاَّ أُمَمٌ أَمْثَالُكُم

There is no animal on the earth, nor any bird that wings its flight,

but is a community like you.

(Qur'an 6: 38)

$$\text{لَخَلْقُ السَّمَاوَاتِ وَالأَرْضِ أَكْبَرُ مِنْ خَلْقِ النَّاسِ وَلَكِنَّ أَكْثَرَ النَّاسِ لَا يَعْلَمُونَ}$$

The creation of the heavens and the earth

is greater than the creation of humankind,

but most of humankind do not know it.

(Qur'an 40: 57)

2.7   We recognize that we are accountable for all our actions:

$$\text{فَمَن يَعْمَلْ مِثْقَالَ ذَرَّةٍ خَيْرًا يَرَهُ}$$

$$\text{وَمَن يَعْمَلْ مِثْقَالَ ذَرَّةٍ شَرًّا يَرَهُ}$$

Then whoever has done an atom's weight of good, shall see it,

and whoever has done an atom's weight of evil, shall see it.

(Qur'an 99: 7–8)

2.8   In view of these considerations we affirm that our responsibility as Muslims is to act according to the example of the Prophet Muhammad (God's peace and blessings be upon him), who:

- Declared and protected the rights of all living beings, outlawed the custom of burying infant girls alive, prohibited wanton killing of living beings for sport, guided his companions to conserve water even in washing for prayer, forbade the felling of trees in the desert, ordered a man who had taken some nestlings from their nest to return them to their mother, and when he came upon a man who had lit a fire on an anthill, commanded, 'Put it out, put it out!';
- Established inviolable zones (*harams*) around Makkah and al-Madinah, within which native plants may not be felled or cut and wild animals may not be hunted or disturbed;
- Established protected areas (*himās*) for the conservation and sustainable use of rangelands, plant cover, and wildlife;
- Lived a frugal life, free of excess, waste, and ostentation;
- Renewed and recycled his meagre possessions by repairing or giving them away;
- Ate simple, healthy food, which only occasionally included meat;
- Took delight in the created world; and
- Was, in the words of the Qur'an, 'a mercy to all beings.'

WE CALL

3.1   We call upon the Conference of the Parties (COP) to the United Nations Framework Convention on Climate Change (UNFCCC) and the Meeting of the Parties (MOP) to the Kyoto Protocol taking place in Paris this December, 2015, to bring their discussions to an equitable and binding conclusion, bearing in mind:

- The scientific consensus on global climate change, which is to stabilize greenhouse gas concentration in the atmosphere at a level that would prevent dangerous anthropogenic interference with the climate systems;
- The need to set clear targets and monitoring systems;
- The dire consequences to the planet Earth if we do not do so;
- The enormous responsibility the COP shoulders on behalf of the rest of humanity, including leading us to a new way of relating to God's Earth.

3.2   We particularly call on the well-off nations and oil-producing states to:

- Lead the way in phasing out their greenhouse gas emissions as early as possible and no later than the middle of the century;
- Provide generous financial and technical support to the less well-off to achieve a phase-out of greenhouse gasses as early as possible;
- Recognize the moral obligation to reduce consumption so that the poor may benefit from what is left of the earth's non-renewable resources;
- Stay within the '2 degree' limit, or, preferably, within the '1.5 degree' limit, bearing in mind that two-thirds of the earth's proven fossil fuel reserves remain in the ground;
- Re-focus their concerns from unethical profit from the environment, to preserving it and elevating the condition of the world's poor.
- Invest in the creation of a green economy.

3.3   We call on the people of all nations and their leaders to:

- Aim to phase out greenhouse gas emissions as soon as possible in order to stabilize greenhouse gas concentrations in the atmosphere;
- Commit themselves to 100 per cent renewable energy and/or a zero emissions strategy as early as possible, to mitigate the environmental impact of their activities;
- Invest in decentralized renewable energy, which is the best way to reduce poverty and achieve sustainable development;
- Realize that to chase after unlimited economic growth on a planet that is finite and already overloaded is not viable. Growth must be

pursued wisely and in moderation; placing a priority on increasing the resilience of all, and especially the most vulnerable, to the climate change impacts already underway and expected to continue for many years to come.

- Set in motion a fresh model of wellbeing, based on an alternative to the current financial model, which depletes resources, degrades the environment, and deepens inequality.
- Prioritize adaptation efforts with appropriate support to the vulnerable countries with the least capacity to adapt, and to vulnerable groups, including indigenous peoples, women, and children.

3.4   We call upon corporations, finance, and the business sector to:

- Shoulder the consequences of their profit-making activities, and take a visibly more active role in reducing their carbon footprint and other forms of impact upon the natural environment;
- In order to mitigate the environmental impact of their activities, commit themselves to 100 per cent renewable energy and/or a zero emissions strategy as early as possible and shift investments into renewable energy;
- Change from the current business model, which is based on an unsustainable escalating economy, and adopt a circular economy that is wholly sustainable;
- Pay more heed to social and ecological responsibilities, particularly to the extent that they extract and utilize scarce resources;
- Assist in the divestment from the fossil fuel driven economy and the scaling up of renewable energy and other ecological alternatives.

3.5   We call on all groups to join us in collaboration, co-operation, and friendly competition in this endeavour, and we welcome the significant contributions taken by other faiths, as we can all be winners in this race:

وَلَكِن لِيَبْلُوَكُمْ فِي مَا آتَاكُم فَاسْتَبِقُوا الْخَيْرَاتِ

But that He (God) may try you in that which He has

given you: So vie with one another

in doing good deeds.

(Qur'an 5: 48)

If we each offer the best of our respective traditions, we may yet see a way through our difficulties.

3.6   Finally, we call on all Muslims wherever they may be:

> Heads of state
> Political leaders
> Business community
> UNFCCC delegates
> Religious leaders and scholars
> Mosque congregations
> Islamic endowments (*awqāf*)
> Educators and educational institutions
> Community leaders
> Civil society activists
> Non-governmental organisations
> Communicators and media

to tackle habits, mindsets, and the root causes of climate change, environmental degradation, and the loss of biodiversity in their particular spheres of influence, following the example of the Prophet Muhammad (peace and blessings be upon him), and bring about a resolution to the challenges that now face us. Allah says in the Qur'an:

وَلاَ تَمْشِ فِي الأَرْضِ مَرَحًا إِنَّكَ لَن تَخْرِقَ الأَرْضَ وَلَن تَبْلُغَ الْجِبَالَ طُولاً

Do not strut arrogantly on the earth.

You will never split the earth apart

nor will you ever rival the mountains in stature.

(Qur'an 17: 37)

We bear in mind the words of our Prophet (peace and blessings be upon him):

'The world is sweet and verdant, and verily Allah has made you stewards in it,

and He sees how you acquit yourselves.'

(Hadith related by Muslim from Abu Saʻid al-Khudri)

LIST OF ACRONYMS

ARC   Alliance of Religions and Conservation

BCE   Before the Common Era

CCMJ   Christian Council for Monetary Justice

CE   Common Era

CFC   Chlorofluorocarbon

COP   Conference of Parties

FBO   Faith Based Organization

FORE   Forum on Religion and Ecology

GDP   Gross Domestic Product

GHG   Greenhouse Gas

GNH   Gross National Happiness

GNP   Gross National Product

ICOREC   International Consultancy on Religion, Education and Culture

IFEES   Islamic Foundation for Ecology and Environmental Sciences

IIED   International Institute for Environment and Development

IMF   International Monetary Fund

IPCC   Intergovernmental Panel on Climate Change

IUCN	International Union for the Conservation of Nature (Union mondiale pour la nature)
LHR	Large Hadron Collider
MAD	Mutually Assured Destruction
MOA	Mokichi Okada Association
NGO	Non-governmental Organization
ODLRC	Ohito Declaration on Religions, Land and Conservation
OECD	Organization for Economic Co-operation and Development
POPs	Persistent organic pollutants
SWC	Saudi Wildlife Commission
UNCED	United Nations Conference on Environment and Development
UNDP	United Nations Development Programme
UNEP	United Nations Environment Programme
UNESCO	United Nations Educational, Social and Cultural Organization
UNFCCC	United Nations Framework Convention on Climate Change
UN MDG	United Nations Millennium Development Goals
WMO	World Meteorological Organization
WWF	World Wide Fund for Nature
WTO	World Trade Organization

GLOSSARY OF ARABIC TERMS[1]

'Adl	–	justice, fairness, equitableness, the mean between excess and falling short.
Al-jabr	–	Algebra
Amānah	–	trust, moral responsibility, honesty.
Awqāf	–	pious foundation (plural *waqf*).
Āyah	–	a verse in the Qur'an, sign (plural *āyāt*).
Dīn	–	the life transaction.
Dīwān	–	a collection of poems (has numerous connotations depending on context).
Fitrah	–	the first nature, the natural, primal condition of mankind in harmony with nature.
Fiqh	–	the science of the application of the Shariah.
Fiqh al-bi'ah	–	environmental jurisprudence.
Hadith	–	reported speech of the Prophet
Haram	–	sacred precinct, a protected area.
Harīm	–	protected space.
Hawālah	–	A debt transfer system.
Himā	–	a reserve for wildlife and forests prohibited to the public.
Hisbah	–	an institution established for the purpose of promoting good and preventing evil.
Hubb	–	love.
Idkhir	–	a cooking herb found in the Hijaz, the region along the western seaboard of Arabia in which Makkah and Madinah are situated.

Ihsān	–	absolute sincerity to Allah in oneself. It is to worship Allah as though you were seeing Him because He sees you.
ihyā' al-mawāt	–	'revival of dead land', bringing wasteland into cultivation.
Ijārah	–	land leased for its usufruct, lease or hire.
'Ilmu'l-khalq	–	Knowledge of Creation.
Īmān	–	belief, faith, acceptance of Allah and His Messenger.
Iqtā'	–	land grants.
Iqtisād	–	moderation, adopting a middle course, being frugal.
Islaām	–	submission to the will of Allah.
Kalām	–	Islamic theology
Khalīfah	–	variously translated as successor, inheritor, heir, viceroy, vicegerent, deputy.
Khushu'	–	humility.
Mīzān	–	balance, scale. Symbol of harmony in creation and also the scales of the Final Reckoning,
Muhājirūn	–	the Companions of Prophet Muhammed who migrated from Madinah to Makkah.
Muhtasib	–	public functionality who supervises markets.
Najm	–	lends itself to be translated both as plants or stars.
Ni'mah	–	gift.
Qādī	–	Islamic judge
Qasīdah	–	ode, poem.
Ribā	–	usury: obtaining something for nothing through exploitation.
Shariah	–	the religious law forming part of the Islamic tradition. It is derived from the religious precepts of Islam, particularly the Qur'an and the Hadith.
Shūrah	–	consultation.
Sunnah	–	the practice of Prophet Muhammad and the first generation of Muslims.
Tawhīd	–	the doctrine of the divine unity.
Zakat	–	a wealth tax paid on certain forms of wealth: gold, silver, staple crops, livestock and trading goods. One of the five pillars of Islam.

---

1 Assistance in compiling this list was sought from Aisha Bewley, *Glossary of Islamic Terms* (London: Ta Ha Publishers, 1998).

LIST OF QUOTATIONS FROM THE QUR'AN

<u>Dedication page</u>

(51: 20–21)   *There are signs on the Earth for people with certainty.*

                   And in your selves as well. Do you not then see?

<u>Introduction</u>

(15: 19–22)   *As for the Earth We have spread it out wide*

                   And set upon it firmly embedded mountains

                   And made everything grow there in balance.

                   And We have provided means of sustenance for you

                   And for all those creatures who do not depend on you.

                   There is not a thing whose storehouses are not with Us

                   And We only send it down in appropriate measure.

                   We send forth pollinating winds

                   And bring down water from the sky for you to drink

                   And you do not control its sources.

(2: 278–279)   *... forego any outstanding dues from usury (interest) ... if you do not then be warned of war from Allah and His Messenger ...*

(40: 57)   *The creation of the heavens and the earth is far greater than the creation of humankind. But most of humankind do not know it.*

Chapter 1

(55: 3–4)   *He (Allah) created man and taught him clear expression.*

(3: 190)   *Surely in the creation of the heavens and earth and in the alteration of night and day there are signs for men possessed of minds.*

(3: 104)   *Let there be a community among you that calls for what is good, urges what is right and forbids what is wrong; those are the ones who have success.*

55: 17)   *The Lord of the two Easts and the Lord of the two Wests*

(24: 40)   *[O]r they are as shadows upon a sea obscure,*
covered by a billow, above which is billow, above which are clouds,
shadows piled one upon another;
when he puts forth his hand, well-nigh he cannot see it

Chapter 2

(55: 8–10)   *Transgress not in the balance (mīzān)*
and weigh with the justice, and skimp not in the balance.
And Earth: He (Allah) set it down for all beings.

(17: 37)   *Do not strut arrogantly on the earth. You will never split the earth apart nor will you ever rival the mountains in stature.*

(24: 39)   *... their works are as a mirage in a spacious plain which the man athirst supposes to be water, till, when he comes to it, he finds it is nothing.*

(21: 32)   *We (Allah) made the sky a well secured canopy – yet from its wonders they turn away.*

(40: 57)   *The creation of the heavens and the earth is far greater than the creation of humankind. But most of humankind do not know it.*

(6: 165)   *... it is He (Allah) who appointed you as stewards on the Earth.*

(55: 6)   *... and the plants (stars) and trees prostrate themselves (to the Creator).*

(24: 45)   *Allah created every animal from water. Some of them creep on their bellies, some that walk on two legs and some on four.*

(6: 38)   *There is no creature crawling on the earth, or those that fly, who are not communities like yourselves.*

(21: 30)   *We (Allah) made every living thing from water.*

(32: 27)   *We (Allah) drive(s) rain to the barren land.*

(30: 41)   *Corruption has appeared in the land and sea, for that men's own hands have earned, that He (Allah) may let them taste some part of that which they have done, that perhaps they may return.*

Chapter 3

(5: 87)   *Do not forbid the good things Allah has made lawful to you. And do not overstep the limits. Allah does not love those who overstep the limits.*

(6: 141)   *It is He (Allah) who produces gardens, both cultivated and wild, and palm trees and crops of diverse kinds, and olives and pomegranates, alike yet different. Eat of their fruits when they bear fruit and pay their due on the day of their harvest. And do not be wasteful for He (Allah) does not love the wasters.*

Chapter 4

(2: 275)   *Those who take* ribā (usury/interest) *will rise up on the Day of Resurrection like someone tormented by Satan's touch.*

Chapter 5

(5: 48)   *Had Allah willed He would have made you a single community. But He wanted to test you regarding what has come to you. So compete with each other in doing good.*

(25: 2)   *He (Allah) created everything and determined them in exact proportions.*

(40: 57)   *The creation of the heavens and the earth is far greater than the creation of humankind. But most of humankind do not know it.*

(51: 20–21)   *There are signs on the Earth for people with certainty.*

                 And in your selves as well. Do you not then see?

(16: 65–69)   *Allah sends down water from the sky,*

                 and by it brings the dead earth back to life.

                 There is certainly a Sign in that for people who hear.

There is instruction for you in cattle.

From the contents of their bellies,

from between dung and blood,

We give you pure milk to drink,

easy for drinkers to swallow.

And from the fruit of the date palm and the grapevine

you derive both intoxicants and wholesome provision.

There is certainly a Sign in that for people who use their intellect.

Your Lord revealed to the bees, [saying]:

'Build dwellings in the mountains and the trees,

And also in the structures which men erect.

Then eat from every kind of fruit

and travel the paths of your Lord,

which have been made easy for you to follow.'

From inside them comes a drink of varying colours,

containing healing for mankind.

There is certainly a Sign in that for people who reflect.

(96: 1–2)   *Read! In the name of your Lord who created; Created man from clots of blood.*

(10: 5–6)   *It is He who appointed the sun to give radiance; and the moon to give light, assigning it in phases ... in the alteration of night and day and what Allah has created in the heavens and earth there are signs for people who have awareness.*

(24: 45)   *Allah created every animal from water. Some of them creep on their bellies, some walk on two legs and some on four. Allah creates whatever He wills.*

(31: 10)   *... and We send down water from the sky and make every generous species grow in it.*

(6: 141)   *It is He who produces gardens, both cultivated and wild, and palm-trees and crops of diverse kinds.*

(30: 30)   *So set your face firmly towards your* dīn *as a pure natural believer*

     Allah originated you in His original creation.

     There is no changing Allah's creation.

     That is the true dīn,

     but most people do not know it.

(67: 3)   *... return thy gaze; seest thou any fissure?*

(42: 5)   *... the heavens well nigh are rent above them.*

(82: 1)   *... when heaven is split open.*

(6: 79)   *I have turned my face to Him who originated (*fatara*) the heavens and the Earth, a pure natural believer.*

(30: 41)   *Corruption has appeared in the land and sea, for that men's own hands have earned, that He (Allah) may let them taste some part of that which they have done, that perhaps they may return.*

(49: 14)   *... the desert Arabs say, 'we have faith', tell them: 'you do not have faith, rather say, "we have submitted"', for faith has not yet entered your hearts.*

(21: 16)   *We did not create heaven and earth and everything between them as a game.*

(24: 39)   *... a mirage in a spacious plain which a man athirst thinks it is water but when he reaches it, he finds it to be nothing ...*

(20: 50)   *Allah gives each thing its created form and then guides it.*

(3: 104)   *Let there be a community among you that calls for what is good, urges what is right and forbids what is wrong; those are the ones who have success.*

(6: 141)   *Eat of their fruits when they bear fruit and pay their due on the day of their harvest. And do not be wasteful. For He (Allah) does not love the wasters.*

(45: 12)   *... be thankful.*

(45: 13)   *He has subjected all that is in the heavens and the earth for your benefit as a gift from Him.*

(38: 27)   *We did not create the heaven and the earth and everything between them to no purpose.*

(5: 48)   *But He (Allah) wanted to test you regarding what has come to you. So compete with each other in doing good.*

(55: 6)   *... and the plants (stars) and trees prostrate themselves (to the Creator).*

6: 38)   *There is no creature crawling on the earth, or those that fly, who are not communities like yourself.*

(55: 60)   *Will the reward for doing good be anything other than good?*

(4: 135)   *... uphold justice and bear witness to Allah, even if it is against yourselves, your parents, or your relatives. Whether they are rich or poor.*

(9: 60)   *Zakat is for the poor, the destitute, those who collect it, reconciling people's hearts, freeing slaves, those in debt, spending in the way of Allah, and travellers. It is ordained by Allah. Allah is All-Knowing, All-Wise.*

(2: 275)   *Those who take usury [ribā] will rise up on the Day of Resurrection like someone tormented by Satan's touch.*

(2: 278–279)   *... and forego any remaining usury if you are a true believer. If you do not know that it is war from Allah and His Messenger.*

(2: 29)   *It is He (Allah) who created everything on the Earth for you.*

(5: 87)   *... do not forbid the good things Allah has permitted you. And do not over step the limits.*

(102: 1–4)   *Fierce competition for this world distracted you until you went down to the graves. No indeed you will soon know! Again no indeed you will soon know!*

(107: 1–3)   *Have you seen him who denies his faith? He is the one who harshly rebuffs the orphans and does not urge the feeding of the poor.*

(42: 38)   *Conduct [your] affairs by mutual consultation.*

(112: 1)   *Say: 'He is Allah, the One, Allah the Eternal.'*

(4: 126)   *What is in the heavens and the earth belongs to Allah: He encompasses*
                everything.

(30: 30)   *Allah originated you in His original creation.*

(55: 1–8)   *The All-Merciful taught the Qur'an,*

                He created man and taught him clear expression.

                The sun and moon both run with precision,

                the plants (stars) and the trees all bow down in prostration,

                He erected heaven and established the balance ...

                so that you would not exceed the balance.

(16: 4)   *He created man from a drop of sperm and yet he is an open challenger.*

(6: 165)   *It is He (Allah) who appointed you as stewards on the earth.*

(33: 72)   *We (Allah) offered the trust to the heavens, the earth and the mountains,*
                but they refused to take it on and shrank from it. But man took it on. He
                is indeed wrong doing and ignorant.

(40: 57)   *The creation of the heavens and the earth is far greater than the creation*
                of humankind.

(55: 3–4)   *He (Allah) created man and taught him clear expression.*

(55: 9–11)   *... weigh with justice and skimp not in the balance. He set the earth down*
                for all beings. With its fruits, its palm trees with clustered sheaths.

(6: 38)   *There is no creature ... who are not communities like yourself.*

(85: 13–14)   *It is He (Allah) who originates and regenerates (nature). He is Ever*
                Forgiving, All-loving.

(25: 63)   *The servants of the Compassionate (Allah) are those who walk the Earth*
                in humility.

(57: 25)   *We (Allah) sent our messengers with clear signs. And sent down the Book*
                and the Balance with them so that mankind might establish justice.

(3: 104)   *Let there be a community among you that calls for what is good, urges*
                what is right and forbids what is wrong; those are the ones who have
                success.

Chapter 6

(30: 41)   *Corruption has appeared in the land and sea, for that men's own hands have earned, that He (Allah) may let them taste some part of that which they have done, that perhaps they may return.*

21: 30)   *We (Allah) made every living thing from water.*

(51: 20–21)   *There are signs on the Earth for people with certainty.*

And in your selves as well. Do you not then see?

(17: 37)   *Do not strut arrogantly on the earth. You will never split the earth apart nor will you ever rival the mountains in stature.*

(2: 275)   *Those who take ribā* (usury/interest) *will rise up on the Day of Resurrection like someone tormented by Satan's touch.*

(3: 104)   *Let there be a community among you that calls for what is good, urges what is right and forbids what is wrong; those are the ones who have success.*

(55: 7–13)   *He (Allah) set up the balance*

so that you may not exceed the balance,

weigh with justice and do not fall short in the balance.

He (Allah) has spread out the earth for living creatures.

With its fruit and date palms with sheathed clusters,

also grain on leafy stems and fragrant herbs.

Then which of the favours of your Lord will you deny?

BIBLIOGRAPHY

Introduction

Bauman, Zygmunt, 'Modernity' in Joel Krieger (ed.), *The Oxford Companion to Politics of the World* (New York: Oxford University Press, 1993).

Gray, John, *Al Qaeda and What It Means to be Modern* (London: Faber and Faber, 2003).

Khalid, Fazlun M, 'Islam and the Environment', in Peter Timmerman and Ted Munn (eds), *Encyclopaedia of Global Environmental Change, volume 5: Social and economic dimensions of global environmental change* (Chichester: John Wiley & Sons, 2002).

Orr, Abdalhalim and Abdassamad Clarke (eds), *Banking: The Root Causes of the Injustices of Our Time*, revised edition (London: Diwan Press, 2009).

Stiglitz, Joseph, *Globalization and its Discontents* (London: Penguin Books, 2002).

Chapter 1

Adorno, Theodore W and Max Horkheimer, *Dialectic of the Environment* (London: Verso, 1997).

Ames, Glenn J, *The Globe Encompassed: The Age of European Discovery, 1500–1700* (Upper Saddle River, NJ: Pearson Prentice Hall, 2008).

Arberry, Arthur J, *Revelation and Reason in Islam* (London: George Allen and Unwin, 1957).

Bernstein, Peter L, *The Power of Gold: The History of an Obsession* (New York: John Wiley and Sons, 2000).

Corradetti, Claudio, 'The Frankfurt School and Critical Theory' (Internet Encyclopaedia of Philosophy (IEP), undated). Available at: http://www.iep.utm.edu/frankfur/

Cromer, the Earl of, *Modern Egypt*, two volumes (London: Macmillan, 1908).

Davies, Glyn, *A History of Money: From Ancient Times to the Present Day* (Cardiff: University of Wales Press, 1994).

Faruqi, Ismail and Lois Faruqi, *The Cultural Atlas of Islam* (New York: Macmillan, 1986).

Flew, Antony (ed.), *The Dictionary of Philosophy* (London: Pan Books, 1979).

Gray, John, *Heresies: Against Progress and Other Illusions* (London: Granta Books, 2004).

Hamilton, Robert, *Earth Dream: The Marriage of Reason and Intuition* (Bideford: Green Books, 1990).

al-Hassani, Salim TS (ed.), *1001 Inventions: The Enduring Legacy of Muslim Civilization*, third edition (Washington, DC: National Geographic, 2012).

Hill, Donald R., *Islamic Science and Engineering* (Edinburgh: Edinburgh University Press, 1993).

Honderich, Ted (ed.), *The Oxford Companion to Philosophy* (Oxford: Oxford University Press, 1995).

King, Mervyn, *The End of Alchemy: Money, Banking and the Future of the Global Economy* (London: Little Brown, 2016).

Kurtzman, Joel, 'Money's Demise', *World Business Academy Perspectives*, 8/2 (1994).

Masood, Ehsan, *Science and Islam: A History* (London: Icon Books, 2017).

Murphy, Brendan, 'Finance: The Unifying Theme' (*The Atlantic*, July 1993). Available at: https://www.theatlantic.com/magazine/archive/1993/07/finance-the-unifying-theme/305148/.

Newsinger, John, *The Blood Never Dried: A People's History of the British Empire* (London: Bookmarks Publication, 2006).

Orr, David, 'What Is Education For? Six myths about the foundations of modern education, and six new principles to replace them', *In Context: The Learning Revolution*, 27 (Winter 1992). Available at: http://www.context.org/iclib/ic27/orr/.

Peet, Richard, *Unholy Trinity: The IMF, World Bank and WTO* (Kuala Lumpur: Strategic Information Research Development, 2003).

Polanyi, Karl, *The Great Transformation: The Political and Economic Origins of Our Time* (Boston, MA: Beacon Press, 2001).

Prideaux, Margi, *Global Environmental Governance, Civil Society and Wildlife: Birdsong After the Storm* (Abingdon: Routledge, 2017).

Quadir, Tarik M, *Traditional Islamic Environmentalism: The Vision of Seyyed Hossein Nasr* (Lanham, MD: University Press of America, 2013).

Quigley, Carroll, *Tragedy and Hope: A History of the World in our Time* (New York: Macmillan, 1966).

Quigley, Carroll, *The Evolution of Civilizations: An Introduction to Historical Analysis* (Carmel, IN: Liberty Fund, 1979).

Rapoport, Yossef and Shahab Ahmed (eds), *Ibn Taymiyya and his Times* (Karachi: Oxford University Press, 2010).

Roberts, JM, *The Pelican History of the World* (London: Pelican Books, 1980).

Rowe, Dorothy, *The Real Meaning of Money* (London: Harper Collins, 1997).

Ryan-Collins, Josh, Tony Greenham, Richard Werner and Andrew Jackson, *Where Does Money Come From? – A guide to the UK monetary and banking system*, second edition (London: New Economics Foundation, 2012).

Saul, John Ralston, *Voltaire's Bastards: The Dictatorship of Reason in the West* (New York: Vintage Books, 1993).

Seabrook, Jeremy, *The Myth of the Market: Promises and Illusions* (Bideford: Green Books, 1990).

Setiono, Benny G, *Tionghoa dalam Pusaran Politik* [*Indonesia's Chinese Community under Political Turmoil*] (Jakarta: TransMedia Pustaka, 2008).

Smith, Adam, *Wealth of Nations* (Ware: Wordsworth Editions Ltd., 2012).

Tarnas, Richard, *The Passion of the Western Mind: Understanding the Ideas That Have Shaped Our Worldview* (London: Pimlico, 1996).

Touraine, Alain, *Critique of Modernity* (Oxford: Basil Blackwell 1995).

Wallin, Nils-Bertil, 'The History of Zero: How was zero discovered?' (YaleGlobal Online, 19 November 2002). Available at: http://yaleglobal.yale.edu/history-zero.

White, Jr, Lynn, 'The Historical Roots of Our Ecologic Crisis', *Science*, 155/3,767 (10 Mar 1967), pp. 1203–1207. Available at: http://www.rci.rutgers.edu/~hallman/PDF/Roots.pdf.

Chapter 2

de Almeida, Miguel Ozorio, *Environment and development: The Founex report on development and environment* (New York: Carnegie Endowment for International Peace, 1971).

Barney, Gerald O, *Global 2000: Entering the Twenty-First Century* (Arlington, VA: Seven Locks Press, 1980).

Brown, Robert W, 'Ancient Civilisation to 300 BC. Introduction: The Invention and Diffusion of Civilisation' (University of North Carolina, 2006).

Carper, Jean, *100 Simple Things You Can Do To Prevent Alzheimer's and Age-related Memory Loss* (London: Vermilion, 2011).

Carson, Rachel, *Silent Spring* (New York: Houghton Mifflin, 1962).

Ceballos, Gerardo, Paul R Ehrlich and Rodolfo Dirzo, 'Biological annihilation via the ongoing sixth mass extinction signalled by vertebrate population losses and declines', *Proceedings of the National Academy of Sciences*, 114/30 (25 July 2017), pp. E6089–E6096.

Colborn, Theo, Dianne Dumanoski and John Peterson Myers, *Our Stolen Future* (London: Penguin, 1996).

Diamond, Jared, *Collapse: How Societies Choose to Fail or Succeed* (New York: Viking, 2005).

Emerson, Ralph Waldo, 'The Ministry of Nature', in Camille Helminski (ed.), *The Book of Nature: A Sourcebook of Spiritual Perspectives on Nature and the Environment* (Bristol: The Book Foundation, 2006).

Hamilton, Stuart, 'Assessing the Role of Commercial Aquaculture in Displacing Mangrove Forest', *Bulletin of Marine Science*, 89/2 (Virginia Key, FL: University of Miami, Rosenstiel School of Marine and Atmospheric Sciences, April 2013), pp. 585–601(17).

Hallmann, Caspar A, Martin Sorg, Eelke Jongejans, et al., 'More than 75 percent decline over 27 years in total flying insect biomass in protected areas' (*PLOS ONE*, 18 October 2017). Available at: http://journals.plos.org/plosone/article?id=10.1371/journal.pone.0185809.

Helminski Camille (ed), *The Book of Nature - A Sourcebook of Spiritual Perspectives on Nature and the Environment.* (Bristol England, Watsonville California: The Book Foundation, 2006).

Independent Commission on International Development Issues, *North–South: A Programme for Survival* (London: Pan Books, 1980).

Jamieson, Alan J, Tamas Malkocs, Stuart B Piertney, et al., 'Bioaccumulation of persistent organic pollutants in the deepest ocean fauna', *Nature: Ecology & Evolution*, 1, article no. 0051 (2017). Available at: http://www.nature.com/articles/s41559-016-0051.

Juniper, Tony, *Saving Planet Earth: What is destroying the earth and what you can do to help* (London: Collins, 2007).

Kasperson, Jeanne, Roger Kasperson, and BL Turner, *The Aral Sea Basin: A Man-Made Environmental Catastrophe* (Boston: Kluwer Academic Publishers, 1995).

Kwiatkowski, Carol F, Ashley L Bolden, Richard A Liroff, et al., 'Twenty-Five Years of Endocrine Disruption Science: Remembering Theo Colborn', *Environmental Health Perspectives*, 124/9 (September 2016), A151–A154. Available at: https://ehp.niehs.nih.gov/ehp746/.

Landrigan, Philip J, Richard Fuller, Nereus JR Acosta, et al., 'The *Lancet* Commission on pollution and health' (*The Lancet*, 19 October 2017). Available at: http://www.thelancet.com/journals/lancet/article/PIIS0140-6736 (17)32345-0/fulltext.

Leakey, Richard and Roger Lewin, *The Sixth Extinction: Biodiversity and Its Survival* (London: Weidenfeld and Nicolson, 1996).

Lehtonen, Markku, 'The environmental-social interface of sustainable development: capabilities, social capital, institutions', *Ecological Economics*, 49/2 (2004), pp. 199–214.

Nasr, Seyyed Hossein, *Man and Nature: The Spiritual Crisis of Modern Man* (London: Unwin Paperback, 1990).

Pimm, SL, GJ Russell, JL Gittleman and TM Brooks, 'The Future of Biodiversity', *Science*, 269 (1995), pp. 347–350.

Rozati, Rosa, PP Reddy, P Reddanna and Rubina Mujtaba, 'Role of environmental estrogens in the deterioration of male factor fertility', *Fertility and Sterility*, 78/6 (December 2002), pp. 1187–1194. Available at: http://www.sciencedirect.com/science/article/pii/S0015028202043893.

World Commission on Environment and Development, *Our Common Future* (Oxford: Oxford University Press, 1987).

Chapter 3

Bell, Harry CP, *The Maldive Islands: An Account of the Physical Features, Climate, History, Inhabitants, Productions, and Trade* (Chennai, New Delhi: Asian Educational Services, 2004).

Braun, Ernest, *Futile Progress: Technology's Empty Promise* (London: Earthscan, 1995).

Daily, Gretchen C and Paul R Ehrlich, 'Population, Sustainability, and Earth's Carrying Capacity', *BioScience*, 42/10 (November 1992).

Diamond, Jared, *Collapse: How Societies Choose to Fail or Succeed* (London: Penguin Books, 2011).

Diffie, Bailey W and George D Winius, *Foundations of the Portuguese Empire, 1415–1850: Europe and the World in the Age of Expansion*, volume 1 (Minneapolis: University of Minnesota Press, 1977).

Galbraith, John Kenneth, *The Culture of Contentment* (London: Penguin, 1993).

Goulet, Denis, *Development Ethics: A Guide to Theory and Practice* (London: Zed Books, 1995).

Gray, Albert and Harry CP Bell (trans and eds), *The Voyage of François Pyrard of Laval to the East Indies, the Maldives, the Moluccas, and Brazil*, two volumes (London: Haklutt Society, 1887).

Gray, John, *Heresies: Against Progress and Other Illusions* (London: Granta Books, 2004).

Hodson, HV, *The Diseconomies of Growth* (New York: Ballantine Books, 1972).

Independent Commission on International Development Issues, *North–South: A Programme for Survival* (London: Pan Books, 1980).

Jackson, Tim, *Prosperity without Growth* (London: Earthscan, 2009).

Layard, Richard, *Happiness: Lessons from a New Science* (London: Penguin Books, 2005).

Meadows, Donella, Jorgen Randers and Dennis Meadows, *Limits to Growth: The 30-Year Update* (London: Earthscan, 2005).

Peiser, Benny, 'From Genocide to Ecocide: The Rape of Rapa Nui', *Energy and Environment*, 16/3–4 (2005), pp. 513–539.

Pomplun, Trent, *Jesuit on the Roof of the World: Ippolito Desiden's Mission to Eighteenth-Century Tibet* (Oxford: Oxford University Press, 2009).

Ponting, Clive, *A New Green History of the World: The Environment and the Collapse of Great Civilisations* (London: Vintage Books, 2007).

Richardson, Brian W, *Longitude and Empire: How Captain Cook's Voyages Changed the World* (Toronto: UBC Press, 2005).

Rist, Gilbert, *The History of Development: From Western Origins to Global Faith* (London: Zed Books, 1997).

Rostow, WW, *The Process of Economic Growth*, second edition (Oxford: Oxford University Press, 1960).

Sachs, Wolfgang (ed.), *The Development Dictionary: A Guide to Knowledge as Power* (London: Zed Books, 1992).

Sachs, Wolfgang (ed.), *Global Ecology: A New Arena of Political Conflict* (London: Zed Books, 1993).

Sen, Amartya, *Development as Freedom* (Oxford: Oxford University Press, 1999).

Simms, Andrew and Victoria Johnson, *Growth isn't possible: Why we need a new economic direction* (London: New Economics Foundation, 2010). Available at: http://b.3cdn.net/nefoundation/f19c45312a905d73c3_rbm6iecku.pdf.

South Commission, The, *The Challenge to the South: The Report of the South Commission* (Oxford: Oxford University Press, 1990).

Spier, Fred, *Big History and the Future of Humanity* (Chichester: Wiley Blackwell, 2015).

United Nations Development Programme, *Human Development Report 2016: Human Development for Everyone* (UNDP, 2016).

Ura, Karma, Sabina Alkire, Tshoki Zangmo, Karma Wangdi, *A Short Guide to Gross National Happiness Index* (Thimphu: The Centre for Bhutan Studies, 2012).

World Commission on Environment and Development, *Our Common Future* (Oxford: Oxford University Press, 1987).

World Economic Studies Division, *World Economic Outlook, October 2017: Seeking Sustainable Growth* (IMF, October 2017).

Wright, Ronald, *A Short History of Progress* (Edinburgh: Canongate, 2005).

Chapter 4

Adams, Henry, *The Degradation of the Democratic Dogma* (New York: Peter Smith, 1949).

Ames, Glenn J, *The Globe Encompassed: The Age of European Discovery, 1500–1700* (Upper Saddle River, NJ: Pearson Prentice Hall, 2008).

Badr, Gamal Moursi, 'Islamic Law: Its Relation to Other Legal Systems', *The American Journal of Comparative Law*, 26/2 (1 April 1978), pp. 187–198.

Banaji, Jairus, 'Islam, the Mediterranean and the Rise of Capitalism', *Journal of Historical Materialism*, 15/1 (2007), pp. 47–74. Available at: http://eprints.soas.ac.uk/15983/1/Islam%20and%20capitalism.pdf.

Benham, Fredric, *Economics: A General Introduction* (London: Pitman, 1960).

Bernstein, Peter L, *The Power of Gold: The History of an Obsession* (New York: John Wiley and Sons, 2000).

Davies, Glyn, *A History of Money: From Ancient Times to the Present Day* (Cardiff: University of Wales Press, 1994).

Diffie, Bailey W and George D Winius, *Foundations of the Portuguese Empire, 1415–1850: Europe and the World in the Age of Expansion*, volume 1 (Minneapolis: University of Minnesota Press, 1977).

Douthwaite, Richard J, *The Ecology of Money* (Totnes: Green Books, 1999).

Graeber, David, *Debt: The First 5,000 Years* (New York: Melville House Publishing, 2011).

Kennedy, Margrit, *Interest and Inflation Free Money* (Seva International, 1995).

Kurtzman, Joel, *The Death of Money: How the Electronic Economy Has Destabilised the World's Markets and Created Financial Chaos* (Toronto: Little Brown & Co., 1993).

Lietaer, Bernard, *The Future of Money: Creating New Wealth, Work and a Wiser World* (London: Century, 2001).

Moore, Karl and David Lewis, *The Origins of Globalization* (New York: Routledge, 2009).

Narayan, MGS, *Calicut: The City of Truth* (Calicut: University of Calicut, 2006).

Orr, Abdalhalim and Abdassamad Clarke (eds), *Banking: The root causes of the injustices of our time*, second edition (London: Diwan Press, 2009).

Piel, Gerard, 'The Acceleration of History', in Ritchie Calder and Hossein Amirsadeghi (eds), *The Future of a Troubled World* (London: Heinemann, 1983).

Penna, Anthony N, *The Human Footprint: A Global Environmental History* (Chichester: Wiley Blackwell, 2015).

Polanyi, Karl, *The Great Transformation: The Political and Economic Origins of Our Time* (Boston, MA: Beacon Press, 2001).

Potter, David Stone, *The Roman Empire at Bay: AD 180–395* (London: Routledge, 2004).

Roberts, JM, *The Pelican History of The World* (London: Pelican Books, 1980).

Rostow, WW, *The Process of Economic Growth*, second edition (Oxford: Oxford University Press, 1960).

Smith, Richard L, *Premodern Trade in World History* (London: Routledge, 2009).

Tawney, RH, *Religion and the Rise of Capitalism: A Historical Study* (London: Penguin Books, 1990).

Chapter 5

Arberry, Arthur J, *The Koran Interpreted* (Oxford: Oxford University Press, 1964).

Asvat, Riyad, *Economic Justice and Shari'a in the Islamic State* (Melbourne: Madinah Press, 2011).

Bagader, Abubakr Ahmed, Abdullatif Tawfik El-Chirazi El-Sabbagh, Mohammad As-Sayyid Al-Glayand, Mawil Yousuf Izzi-Deen Samarrai, and Othman Abd-ar-Rahman Llewellyn (eds), 'Environmental Protection in Islam', Environmental Policy Law Paper No. 20, second revised edition (IUCN, Gland Switzerland and Cambridge UK, 1994). Available at http://cmsdata.iucn.org/downloads/eplp_020reven.pdf.

Berry, Thomas, *The Great Work: Our Way into the Future* (New York: Bell Tower/ Random House, 1999).

Bewley, Aisha, *The Glossary of Islamic Terms* (London: Ta-Ha, 1988).

Dockrat, Hashim Ismail, 'Islam, Muslim Society and Environmental Concerns: A Development Model Based on Islam's Organic Society', in Richard C Foltz, Frederick M Denny and Azizan Baharuddin (eds), *Islam and Ecology: A Bestowed Trust* (Cambridge, MA: Harvard University Press, 2003).

Dutton, Yasin, 'The Environmental Crisis of Our Time: A Muslim Response', in Richard C Foltz, Frederick M Denny and Azizan Baharuddin (eds), *Islam and Ecology: A Bestowed Trust* (Cambridge, MA: Harvard University Press, 2003).

Faruqi, Ismail and Lois Faruqi, *The Cultural Atlas of Islam* (New York: Macmillan, 1986).

Foltz, Richard C, Frederick M Denny and Azizan Baharuddin (eds), *Islam and Ecology: A Bestowed Trust* (Cambridge, MA: Harvard University Press, 2003).

Gray, John, *False Dawn: The Delusions of Global Capitalism* (London: Granta, 2009).

Grim, John and Mary Evelyn Tucker, *Ecology and Religion* (Washington, DC: Island Press, 2014).

al-Habib, Shaykh Muhammad ibn, *The Diwan* (Cape Town: Madinah Press, 2001).

Izzi Dien, Mawil, *The Environmental Dimensions of Islam* (Cambridge: The Lutterworth Press, 2000).

Johnson-Davies, Denys (trans.), *The Island of Animals* (London: Quartet Books, 1994).

Kassis, Hanna E, *A Concordance of the Qur'an* (Oakland, CA: University of California Press, 1983).

Khalid, Fazlun, 'Guardians of the Natural Order', *Our Planet*, 8/2 (July 1996).

Khalid, Fazlun, 'Islamic Pathways to Ecological Sanity: An Evaluation for the New Millennium. Ecology and Development', *Journal of the Institute of Ecology*, 3 (2000).

Khalid, Fazlun, 'Applying Islamic Environmental Ethics', in Richard Foltz (ed.), *Environmentalism in the Muslim World* (New York: Nova Science, 2005).

Khalid, Fazlun, 'The Environment and Sustainability: An Islamic Perspective', in Colin Bell, Jonathan Chaplin and Robert White (eds), *Living Lightly, Living Faithfully: Religious faiths and the future of sustainability* (Cambridge: Faraday Institute of Science and Religion, 2013).

Khalid, Fazlun, 'Exploring Environmental Ethics in Islam: Insights from the Qur'an and the Practice of Prophet Muhammad' in John Hart (ed.), *Religion and Ecology* (Wiley Blackwell: Oxford, 2017).

Khalid, Fazlun and Joanne O'Brien (eds), *Islam and Ecology* (London: Cassell, 1992).

Khalid, Fazlun and Ali Khamis Thani, *A Teachers Guide Book for Islamic Environmental Education* (Birmingham: IFEES, 2007). Available at: http://www.ifees.org.uk/wp-content/uploads/2015/04/13751866541.pdf.

Khan, Muhammad Akram, *Economic Teachings of Prophet Muhammad* (Delhi: Oriental Publications, 1992).

Kovel, Joel, *The Enemy of Nature: The End of Capitalism or the End of the World?* (London: Zed Books, 2007).

Masri, Al-Hafiz BA, *Animals in Islam* (London: Athene Trust, 1989).

al-Mawardi, Abu'l Hassan and Assadullah Yate (trans.), *Al-Ahkam As Sultaniyyah: The Laws of Islamic Governance* (London: Ta-Ha, 1996).

Muhammad, Ahsin Sakho, Husein Mumammad and Roghib Mabrur et al. (eds), *Fiqih Lingkungan* (Jakarta: Laporan, 2004).

Murata, Sachiko and William C Chittick, *The Vision of Islam: The Foundations of Muslim Faith and Practice* (London: IB Tauris, 1996).

Nasr, Seyyed Hossein, *Man and Nature: The Spiritual Crisis of Modern Man* (London: Unwin Paperback, 1990).

Nasr, Seyyed Hossein, *The Need for a Sacred Science* (Richmond: Curzon Press, 1993).

Özdemir, Ibrahim, 'Towards an Understanding of Environmental Ethics from a Qur'anic Perspective', in Richard C Foltz, Frederick M Denny and Azizan Baharuddin (eds), *Islam and Ecology: A Bestowed Trust* (Cambridge, MA: Harvard University Press, 2003).

Quigley, Carroll, *The Evolution of Civilizations: An Introduction to Historical Analysis* (Carmel, IN: Liberty Fund, 1979).

Sampson, Anthony, *The Money Lenders* (London: Hodder and Stoughton, 1981).

Surty, Muhammad Ibrahim HI, *Towards Understanding Qur'anic Arabic* (Birmingham: QAF, 1993).

Chapter 6

Asara, Viviana, Iago Otero, Federico Demaria and Esteve Corbera, 'Socially sustainable degrowth as a social–ecological transformation: repoliticising sustainability', *Sustainability Science*, 10/3 (July 2015), pp. 375–384.

'Encyclical Letter *Laudato Si*' of the Holy Father Francis on Care for Our Common Home' (Libreria Editrice Vaticana, undated). Available at: http://w2.vatican.va/content/francesco/en/encyclicals/documents/papa-francesco_20150524_enciclica-laudato-si.html.

Lent, Jeremy, *The Patterning Instinct: A Cultural History of Humanity's Search for Meaning* (New York: Prometheus Books, 2017).

Mason, Paul, *Postcapitalism: A Guide to Our Future* (London: Allen Lane, 2015).

Nassar, Lara and Juliet Dryden (ed.), *The Arab Environmental Governance Charter* (West Asia–North Africa Institute, December 2016).

Piketty, Thomas and Arthur Goldhammer (trans.), *Capital in the Twenty-First Century* (Cambridge, MA: Harvard University Press, 2014).

Ponting, Clive, *A New Green History of the World: The Environment and the Collapse of Great Civilisations* (London: Vintage Books, 2007).

Ripple, William J, Christopher Wolf, Thomas M Newsome, et al., 'World Scientists' Warning to Humanity: A Second Notice' *BioScience*, 67/12 (1 December 2017), pp. 1026–1028. Available at: https://academic.oup.com/bioscience/article/67/12/1026/4605229.

Simms, Andrew and Victoria Johnson, *Growth isn't possible: Why we need a new economic direction* (London: New Economics Foundation, 2010).

Spratt, Stephen and Mary Murphy, *The Great Transition: A tale of how it turned out right* (London: New Economics Foundation, 2009).

Toffler, Alvin, *Future Shock* (New York: Pan Books, 1971).

Tomlinson, John, *Honest Money: A Challenge to Banking* (Deddington, Oxon.: Helix, 1993).

Vadillo, Umar Ibrahim, *The Return of the Gold Dinar: A study of money in Islamic law and the architecture of the gold economy* (Kuala Lumpur: Madinah Press, 2004).

INDEX

Note: Page numbers followed by n denotes notes.